FROM ORWELL'S PROBLEM TO PLATO'S PROBLEM

KNOWLEDGE AND LANGUAGE

Volume I

From Orwell's Problem to Plato's Problem

Edited by

ERIC REULAND

Department of Linguistics, University of Utrecht,
The Netherlands

and

WERNER ABRAHAM

Department of German, University of Groningen,
The Netherlands

KLUWER ACADEMIC PUBLISHERS

DORDRECHT / BOSTON / LONDON

Library of Congress Cataloging-in-Publication Data

Knowledge and language / edited by Eric Reuland and Werner Abraham.
      p.   cm.
    Papers from a conference held May 21-25, 1989 on the occasion of
  the 375th anniversary of the University of Groningen.
    Includes bibliographical references and indexes.
    Contents: v. 1. From Orwell's problem to Plato's problem -- v.
  2. Lexical and conceptual structure
    ISBN 0-7923-1789-0 (v. 1 : alk. paper). -- ISBN 0-7923-1790-4 (v.
  2 : alk. paper)
    1. Language and languages--Congresses. 2. Linguistics-
  -Congresses. 3. Knowledge, Theory of--Congresses. I. Reuland,
  Eric J. II. Abraham, Werner.
  P23.K57 1992
  410--dc20   92-14226

ISBN 0-7923-1789-0
ISBN 0-7923-1888-9 (Set)

Published by Kluwer Academic Publishers,
P.O. Box 17, 3300 AA Dordrecht, The Netherlands.

Kluwer Academic Publishers incorporates the publishing programmes
of D. Reidel, Martinus Nijhoff, Dr W. Junk and MTP Press.

Sold and distributed in the U.S.A. and Canada
by Kluwer Academic Publishers,
101 Philip Drive, Norwell, MA 02061, U.S.A.

In all other countries, sold and distributed
by Kluwer Academic Publishers Group,
P.O. Box 322, 3300 AH Dordrecht, The Netherlands.

Printed on acid-free paper

Printed in The Netherlands

TABLE OF CONTENTS

ACKNOWLEDGEMENTS

At this occasion the editors would like to thank all those involved in the organization of the Conference on Knowledge and Language. Without the help and enthusiasm of many members, both staff and students, of the departments of General Linguistics, General and Comparative Literature, and History, of Groningen University, this conference, and hence this publication would not have been possible. In particular we would like to thank the other members of the organizing committee, Jan Koster and Henny Zondervan. In all matters of organization and planning, Liesbeth van der Velden provided invaluable help, and so did Marijke Wubbolts. We would also like to acknowledge the cooperation of the then Dutch Defense Minister Frits Bolkestein and his staff in the organization of the public debate with Noam Chomsky on The Manufacture of Consent, as well as the contributions by the chairmen and panel members.

The conference was characterized by lively and fundamental discussions. At this point we would like to thank those who contributed to that atmosphere by their presentations, and who for various reasons could not submit their contribution for publication, notably, Manfred Bierwisch, Denis Bouchard, Melissa Bowerman, Gisbert Fanselow, Sascha Felix, Johan Galtung, Alessandra Giorgi, Giuseppe Longobardi, David Pesetsky, Dan Sperber, Michael Tanenhaus and Hayden White.

We also wish to express our gratitude for the financial support by grants received from Kluwer Academic Publishers, Wolters-Noordhoff Publishing Company, NCR Handelsblad, the Royal Dutch Academy of Sciences, the British Council in The Netherlands, the Stichting H.S. Kammingafonds, the Stichting Groninger Universiteitsfonds, the Faculty of Arts and the Executive Board of Groningen University.

We are also very grateful to the anonymous reviewers for encouraging comments. Finally, we thank Kluwer Academic Publishers, especially Irene van den Reydt and Martin Scrivener, for their enthusiasm and support during the preparation stage of these volumes.

The Editors

ERIC REULAND AND WERNER ABRAHAM

INTRODUCTION

1. BACKGROUND

This volume is one of three which emerged from the Conference on Knowledge and Language, held from May 21—May 25, 1989, at the occasion of the 375th anniversary of the University of Groningen.

Studying the relation between knowledge and language, one may distinguish two different lines of inquiry, one focussing on language as a body of knowledge, the other on language as a vehicle of knowledge.

Approaching language as a body of knowledge one faces questions concerning its structure, and the relation with other types of knowledge. One will ask, then, how language is acquired and to what extent the acquisition of language and the structure of the language faculty model relevant aspects of other cognitive capacities.

If language is approached as a vehicle for knowledge, the question comes up what enables linguistic entities to represent facts about the world. To what extent does this rely on conventional aspects of meaning? Is it possible for language, when used non-conventionally as in metaphors, to convey intersubjective knowledge? If so (and it does seem to be the case), one may wonder what makes this possible.

The aim of this conference was to investigate the role of conceptual structure in cognitive processes, exploring it from the perspectives of philosophy of language, linguistics, political philosophy, psychology, literary theory, aesthetics, and the philosophy of science.

The themes of these three volumes reflect the themes of the conference.

2. THREE THEMES AND THEIR CONNECTIONS

In the volume *From Orwell's Problem to Plato's Problem* basic issues are discussed concerning the acquisition of linguistic and non-linguistic knowledge, including the relation between the knowledge that is acquired and the evidence giving rise to it.

Plato's problem is how we can know so much even when the evidence available to us is so sparse. Inborn knowledge structures may enhance acquisition, as in the case of the baffling rate at which the child, on the basis of scant evidence, acquires all it needs to know in order to speak its mother tongue. Orwell's problem is why we know and understand so little even when the evidence available to us is so rich. In some domains the problems overlap, since inborn structure may also have the effect of

1

Eric Reuland and Werner Abraham (eds), Knowledge and Language, Volume I, From Orwell's Problem to Plato's Problem: 1—10.
© 1993 *by Kluwer Academic Publishers. Printed in the Netherlands.*

impeding the acquisition of knowledge, as perhaps in the case of human political history, where little seems to be learned and an abundance of evidence apparently does not suffice to stop repetition of identical errors and blunders.

These two problems are connected by the question of how much structure is to be assigned to the mind outside the domain of specific capacities, such as language, vision, etc., and where or how such structure shows up.

We cannot expect to find easy answers to questions of this type. And, clearly, the investigation of linguistic knowledge alone is far from being sufficient. If answers can be found at all this will require the joint effort of many scholars in a wide range of fields. Yet it is not accidental or unjustified that linguistics figures so prominently in the discussion. The recent history of linguistics sets a significant example, in that linguistics has moved away from speculations as to how one might proceed to explain the convergence observed when human beings acquire language, to examining detailed proposals that set out to provide such an explanation. Although, as in any science, our understanding of the processes involved is limited, considerable progress towards solving Plato's problem in the domain of linguistic knowledge has been made, and its achievements may provide one with a standard of precision for the characterization of knowledge structures in general.

The volume on *Lexical and Conceptual Structure* addresses the nature of the interface between conceptual and linguistic structure.

It is a long-standing observation that there is a correspondence between formal linguistic and notional categories, which is incomplete, yet real. Presumably every language will contain a formal category which at least contains the expressions denoting prototypical objects, another containing the expressions which denote prototypical actions, etc. This does not affect the autonomy of linguistic and cognitive structure. Cognitive notions like "object" or "action" have linguistic constructs such as N, or V as their structural reflexes (Canonical Structural Realizations, using the term of Grimshaw (1981)). Also a substantive relation such as selection is structurally reflected as subcategorization.

If formal linguistic categories have images at some level of cognitive structure, the question arises what the nature of this level is; what are its elements, how do they combine, etc.? What is the relation between "cognitive structure" and the domain of interpretation? Finding the answer to questions of this type is of the utmost importance in order to understand how the language faculty relates to the other cognitive abilities.

One of the important challenges facing us is to get a substantive theory of the relation between meanings as linguistic entities and real world entities off the ground.

Contrary to what one might think, the phenomenon of metaphor is important for this enterprise. By its very nature, a metaphor is non-conventional. Hence, any theory based on the idea that there are fixed links between minimal objects of linguistic structure and concepts anchored in reality will have to treat metaphor as marginal, and cannot explain that people use metaphors and understand them. Relegating this issue to a theory of language use does not help solve it. In the absence of fixed linguistic conventions to resort to, the only avenue to exploit appears to be invoking shared underlying schemata for analyzing and categorizing the world. Metaphor may in fact constitute one of the best windows for investigating the conceptual structure and its underlying principles since in this case the linguistic structure itself does not interfere with the entities observed. Let these considerations suffice to identify a possible program for empirical research.

The volume on *Metaphor* addresses issues in this program. It concerns the cognitive status of figurative and metaphoric use of language. Metaphor has the air of paradox. It is generally understood as not rule governed but free. But, if it is free from rules, how is it possible for people to converge in their interpretations? It might seem that only a common conceptual structure, limiting the range of interpretations would facilitate this, but this leaves open the question of how precisely interpretation proceeds. Metaphoric language is often associated with improperness, or at least imprecision. Yet, it has an important function in scientific texts. How can figurative language be true or false? How is it possible that metaphor is often used rather to enhance precision. Again, this seems to require an independent conceptual structure.

3. THE STRUCTURE OF THIS VOLUME

We will now sketch the structure of this volume, summarizing the contents of the 10 chapters, and indicating their connections.[1]

Like the conference, this volume starts out with a discussion of Plato's and Orwell's problems, and how they are instantiated in the social-political domain.[2] How can it be the case that so often beliefs are 'firmly held and widely accepted' that are 'completely without foundation and often plainly at variance with obvious facts about the world'? What is the role of the media in this state of affairs? What is the role of factors like shared concepts and values, and group formation and group solidarity? In the first chapter, which is based on material circulated among the participants before the conference (the Preliminary statement referred to in Chomsky's contribution) and on an introduction to the topic presented at the conference, Reuland, takes the characterization of Orwell's problem in Chomsky (1986) as a starting point. Comparing the position of the media in the USA as described and analyzed in Herman and Chomsky (1988) with the position of the media in Western Europe, in particular The Netherlands, a number of differences are noted, and evaluated. It is suggested that also in the study of Orwell's problem, the distinction between conscious and unconscious knowledge should be observed. Thus, part of the problem of "why we know and understand so little even when the evidence available to us is so rich", in fact instantiates Plato's problem. The chapter ends with the suggestion that it is conceivable that there exists a "social grammar", like linguistic grammar in part determined by genetic endowment.

Chomsky's contribution in chapter 2, considers the nature of the mind, asking how "worlds are made" by its workings. Its aim is to assess where we stand today, and where we might hope to advance, in pursuing a number of crucial questions of contemporary culture. The "cognitive revolution" of the past generation shifted the focus of inquiry from behavior and its products to the inner mechanisms of the mind/brain that enter into action and interpretation. In some domains (notably, language and vision), computational-representational theories of mind were developed with some success. The mental mechanisms postulated in these inquiries relate to other elements of the world in several ways: they grow along an internally-directed course under the effect of the external environment; they may have an intensional character, being "about" something else to which they refer or otherwise relate; they are used in social interchange, including communication and its other face, indoctrination and control. In making worlds, people may seek to enhance or to prevent understanding, and there may be many factors, individual and social, entering into such choices.

Illuminating parallels between conclusions that were reached during the intellectual and social upheavals of the 17th century and issues that are relevant today are discussed. In a few domains, there has been real progress; others remain obscure, as is shown — among them, the determiners of the mental construction of social worlds. Progress has sometimes led towards an explanatory theory of a certain depth, in particular, in connection with some aspects of "Plato's problem"; in other cases, it is argued, quite simple models are well-confirmed, notably the variant of "Orwell's problem" found in social systems lacking powerful state controls.

In chapter 3, Bracken uncovers a parallel between contemporary sceptical thinking and the debates conducted at the times of Sextus Empiricus and Descartes. In the century before Descartes, Protestants already seek to leave to each individual the discernment of truth independently of any intermediary agency such as the Church. To weaken the Protestant cause, counter-reformers disseminate the sceptical arguments of Sextus Empiricus. But Descartes sees that such sceptical arguments also undermine all science. In the end, Descartes grounds knowledge in the innate ideas present as part of each human's nature rather than upon anything abstracted from empirical data. In that way the core of knowledge is not 'externally' mediated.

A sceptical crisis somewhat like Descartes' is encountered by Orwell's Winston in *1984*, this time engendered by political forces. Orwell, like Descartes, it is argued, sees that the aim of the ultimate 'power trip' is absolute control of the mind. That is why, again like Descartes, he seeks our salvation in unmediated primitive truths.

An appeal to a highly structured human nature as a basis for knowledge in which all humans share, is still not popular. That appeal can be seen as a response to a new sceptical crisis, one which has permeated much post-World War II philosophical thought. In the end, neither the methods nor the goals of the counter-reformers appear to have changed over the past three centuries.

Other contributions focus on Plato's problem, approaching it from the point of view of a different cognitive domain, namely language learning, presenting theoretical discussions of foundational issues and in-depth investigations of empirical problems. This is directly relevant to the broader issues, since progress in other areas of Plato's problem (and Orwell's problem) can only be made by investigations at least as detailed as those presented here.

In the domain of language acquisition three main questions can be identified. One concerns the development of the 'language organ'. Are all general linguistic principles available right from birth, or are there qualitative changes in the linguistic principles available to the child during her development, i.e., is language learning (in part) a maturational process? Another question is to what extent strictly linguistic and general cognitive principles must be distinguished, and if so, how they interact. The last question concerns how the data in the input determine the development of the grammar and the choices to be made. All three questions are touched at, in some form or other, in the contributions of chapters 4—9. Nevertheless, reflecting current acquisition research in general, the focus is on the last question: the role of parameter setting and markedness hierarchies in language acquisition.

In chapter 4, Roeper outlines a theory of parameter setting that can address the prior problem of how the primary linguistic data is analyzed and what universal grammar must bring to bear to accomplish it. The chapter presents a detailed discussion of the process of parameter setting in relation with the fact that certain acquisition stages may have copying where adult stages have movement. In particular, Roeper investigates the phenomenon of 'tense copying' in English (*it does fits, he did left*) in the light of recent proposals by Chomsky and Pollock about the relation between verb and inflection, and the status of *do*-insertion. Roeper notes that Chomsky's least effort principle and the full interpretation principle, given the fact that children do not acquire adult grammar instantaneously, may force changes that may not look like simple progressive changes. I.e., either short derivations, or "do-insertion" may appear in children's grammars where they are not required in adult grammar, since certain derivations are not available in the child's grammar. In addition, cases of *wh*-copying in English and German and clitic copying in French are discussed. It is argued that the existence of copying stages is a natural corollary of the Lexical Learning Hypothesis (e.g. Borer 1984). Roeper proposes that copying arises if the parameter (lowering of inflection versus raising of the verb in the case of 'tense copying') has not yet been resolved. Copying provides an overt, monolevel representation of a derivation which otherwise would be systematically ambiguous. By way of conclusion, these results are put into a biological perspective.

It has been pointed out that besides aspects of language structure that lend themselves to parametrization, there are also aspects that do not (cf. Roeper and Williams 1987). This distinction appears to correlate with that between formal and substantive properties, suggesting a bipartite division: *substantive versus formal*, or *conceptual versus grammatical*. It is not unlikely, though, that one more level is involved in what is broadly speaking the 'language faculty', namely an interpretation domain, with its own formal and substantive properties. Results that may bear on this issue are presented in the next chapter.

In chapter 5, Roeper and De Villiers investigate how cognition connects to grammar, and how these are intertwined in the process of acquisition. This contribution focuses on the acquisition of bound variable interpretations. On the one hand, such interpretations depend on the cognitive ability to make pairings of elements, on the other hand, grammar places significant constraints on the availability of bound variable readings.

In a range of studies with preschool children, it was found that bound variables are available quite early. However, the capacity to link this response to grammatical environments does not emerge immediately. Children recognize the contexts in which a bound variable reading is obligatory but, initially, the bound variable response is overgeneralized to include environments where it is excluded in adult grammar by subjacency. This cannot be explained by the unavailability of subjacency at that stage, since in other constructions the child's grammar does obey subjacency. The assumption explored is that the children's behaviour in this domain is, nevertheless, licensed by their grammars. The sequence of acquisition, here, is related to the development of functional categories, in particular specifier nodes. Functional categories are delayed in emergence (cf. Lebeaux (1988) and others). Before specifiers are present, quantificational determiners can be there, but must be treated as adverbs. In addition, projections lacking specifiers (e.g. CPs) do not count as maximal. Together, this ensures that the problematic readings are available without violating subjacency in the child's grammar (although they would in the adult's). It is argued that certain decisions in this domain are taken, despite an inherent indeterminacy in the data. Such decisions must be linked to a unique trigger in the data. Certain acquisition principles that are not visible within synchronic grammar by themselves are needed to guarantee that available data does not mislead children.

These results call into question the role of the subset principle in this domain. Thus, this contribution relates to the discussion in chapters 8 and 9.

The underlying question of Manzini's contribution in chapter 6 is what components of grammar can be parametrized and in which way. This question is approached in the light of the lexical parametrization principle (Borer 1984), according to which all parameters are lexical. From this perspective, a restrictive theory of parameters is nothing but a restrictive theory of grammatical classes, i.e., of categories in the widest sense of the term. Specifically, Manzini investigates the null subject and Verb-to-Infl parameters and their interaction. A syntactic analysis of he various settings of the Verb-to-Infl parameter is presented, which bears directly on the acquisition of the Verb-to-Infl parameter in general and the issues discussed in the previous chapter in particular. The proposal accounts for the various settings found in English, French and Italian. It is proposed that Infl is lexical in English. In the relation between Infl and Verb, Infl is, then, like an expletive element (note that, from this perspective, the tense copying mechanism described by Roeper in chapter 4 is a quite natural process). By contrast, Infl is not lexical in Italian and French. The other parameter is whether Infl is a head or not. If it is not, the language has the null-subject property (Italian); if, however, Infl is a head (French) the language is not null-subject. Intuitively, this reverses the notion of strength, as hitherto construed. From this perspective, Infl in English is strongest, in Italian it is weakest, with strength reflecting the analytic-synthetic opposition.

Much like chapters 4 and 5, though in a different vein, chapter 7, by Jakubowicz, is concerned with the problem of the relation between linguistic theory and language acquisition facts.

It has been observed that at a certain point of development the child's performance on anaphor binding is better than her performance on pronominal binding. In other words, bound prominals occur in environments where they should be ruled out.

Results of two experiments are presented which show that the computational devices underlying the setting up of coreference relations are already operative when children are about 3 years old. The fact that the performance with pronouns is less perfect than that with anaphors is explained under the hypothesis that while Binding Principles are ready to apply from the start of the acquisition process, they must await for the child to have at her command the entire set of anaphors and pronouns of the language to which she is exposed. It is concluded that a continuity-plus-lexical-learning-hypothesis of language acquisition accounts for the facts discussed and should be preferred over alternative proposals on theoretical as well as empirical grounds.

Chomsky (1986) makes explicit an important conceptual shift in defining the object of linguistic inquiry. Often, languages are understood as sets of well-formed expressions; the study of language is, then, thought to be the study of such sets and their properties. Chomsky argues that such a concept of External/Extensional language (E-language) is in fact irrelevant for the goals of linguistic investigation. Instead, the proper object of study is the linguistic knowledge internalized in the mind: the Internal/Intensional language (I-language). This distinction between E-language and I-language plays a key role in a debate conducted in chapters 8 and 9.

In chapter 8, Kapur, Lust, Harbert, and Martohardjono present a critical assessment of the 'Subset Principle', as it has been developed by Wexler and others. They argue that the subset principle as currently used in language acquisition research (in particular, the acquisition of anaphora), should be distinguished from an earlier, more formal, version. The learnability model based on the subset principle is contrasted both formally and empirically with an approach in which markedness hierarchies are directly expressed in the theory of Universal Grammar. It is argued on the basis of crosslinguistic variation in binding domains and crosslinguistic studies of first language acquisition that the binding domains are not parametrized in accordance with the subset principle, which, the authors claim, disconfirms this proposal empirically. Finally, the authors relate their assessment of the subset principle to the E-language/I-language distinction (Chomsky 1986). It is argued that the failure of the subset principle is due to its being an E-language principle, and not a principle of I-language.

In chapter 9, Wexler takes up the challenge raised in chapter 8. First, he shows that the subset principle has an intensional interpretation as well. He, then, argues that the idea behind the subset principle is that, ideally, markedness hierarchies should follow from independent principles, rather than being stipulated as part of the theory of grammar. Therefore, whenever markedness hierarchies are stipulated, this constitutes a loss in explanation. So, ultimately, the program developed in the previous chapter could result in giving up explanation as a goal of linguistic theorizing. Admittedly, facts might force such an outcome. But, surveying the facts of chapter 8, Wexler argues that at least in this domain they do not. Some of those facts are in fact at variance with the markedness approach of chapter 8.

The question of how knowledge is acquired relates to the question of how it is mentally represented and processed. In recent years a discussion has come up about the mental status of symbolic representations and linguistic principles. The discussion started off with the emergence of connectionist models of cognitive processes (see, e.g. McClelland et al. 1986). In these models questions concerning the micro-structure of the brain are connected to the architecture of higher levels of mental organization. The main claim of the connectionist approach is that information is represented in the form of patterns of activation levels of neurons in neural networks. There is no simple relation between these patterns and the discrete elements and rules figuring in standard symbolic representations of knowledge (as for instance in generative grammars). Although it is possible to implement symbolic representations in such systems, it is argued that, if their non-discrete character is taken seriously, non-trivial and more accurate predictions are made about various processes that take place, including the acquisition of knowledge. Rumelhart and McClelland (1986) claim, for instance, that the pattern of acquisition of English past tense verb forms follows from the general properties of the learning system, without recourse to a notion like rule of grammar, or a specific language acquisition device. This approach, then, could have strong implications for our current views on the nature of the Plato's problem. However, Pinker and Prince (1987), examining the proposal in Rumelhart and McClelland (1987) in detail, show that the actual acquisition pattern differs substantially from what connectionist theories predict, and that the problems in fact involve the essence of that approach. (We will not try to summarize this discussion here, but rather refer to the works cited; for extensive general discussion of the issue, see the various contributions in Lycan (1990).)

In his contribution to the present book, Levelt addresses the issue from another angle. In chapter 10, Levelt shows that also access to knowledge bears on the evaluation of connectionist theories. He discusses one of the most central issues in a psycholinguistic theory of the speaker, namely, the management of lexical access. A speaker selects lexical items from a very sizeable lexicon at an average rate of some two to three per second. These items are phonologically encoded at the rate of about fifteen speech sounds per second.

Lexical selection is driven by the speaker's communicative intentions and constrained by the grammatical restrictions that lexical items impose on their linguistic environment. All existing theories of lexical selection incorporate, in one way or another, the notion that the selection of an item depends on the satisfaction of a set of semantic criteria by the concept to be expressed. However, all theories fall in the trap of the "hypernym problem". They do not prevent that all of an item's hypernyms, or superordinates, are also selected if the relevant semantic conditions are met. This problem is solved by instantiating two principles: the principle of specificity and the core principle. These two principles are implemented in a logical (i.e. non-connectionist) network. Phonological encoding only begins upon selection of a lexical item. Experimental evidence shows that, contrary to the claim of current connectionist theories, a merely activated, but not yet selected, lexical item is phonologically inactive. That is, accessing the linguistic knowledge represented in the lexicon during speech production is a two-stage process with a strict succession of lexical selection and phonological encoding. This testifies to the modular organization of lexical access: lexical selection and phonological encoding are modular processes with a minimum of forward or backward activation spreading between them. Thus, these results cast severe doubt on the interpretation of connectionist models as realistic hypotheses about the architecture of the mind.

NOTES

[1] Some of these summaries are based on abstracts provided by the authors, but greatly abbreviated by the editors. Thus, the editors will assume responsibility for all shortcomings.
[2] The discussion of Orwell's problem was essentially conducted on the first conference day, culminating in a debate between Noam Chomsky and the HT Dutch defense minister Frits Bolkestein on the manufacture of consent.

REFERENCES

Borer, Hagit: 1984, *Parametric Syntax*, Foris, Dordrecht.
Borer, Hagit and Kenneth Wexler: 1987, 'The Maturation of Syntax', in T. Roeper and E. Williams (eds.), *Parameter Setting*, Reidel, Dordrecht.
Chomsky, Noam: 1986, *Knowledge of Language: Its Nature, Origin and Use*, Praeger, New York.
Grimshaw, Jane: 1981, 'Form, Function, and the Language Acquisition Device', in C. L. Baker and J. J. McCarthy (eds.), *The Logical Problem of Language Acquisition*, MIT Press, Cambridge, Massachusetts.
Herman, Edward and Noam Chomsky: 1988, *Manufacturing Consent*, Pantheon, New York.
Lebeaux, David: 1988, *Language Acquisition and the Form of Grammar*, dissertation, University of Massachusetts, GLSA, Amherst.
Lycan, William (ed.): 1990, *Mind and Cognition*, Basil Blackwell, Cambridge, Massachusetts.
Manzini, Rita: 1989, 'Categories and Acquisition in the Parameters Perspective', in R. Carston (ed.), *UCL Working Papers in Linguistics, 1*.
McClelland, J., D. Rumelhart and The PDP Research Group: 1986, *Parallel Distributed Processing: Explorations in the Microstructure of Cognition*, Bradford Books/MIT Press, Cambridge, Massachusetts.
Pinker, Steven and Alan Prince: 1987, 'On Language and Connectionism: Analysis of a Parallel Distributed Processing Model of Language Acquisition', Occasional Paper No. 33, Center for Cognitive Science, MIT, Cambridge, Massachusetts.
Roeper, Tom and Edwin Williams: 1987, 'Introduction', in T. Roeper and E. Williams (eds.), *Parameter Setting*, Reidel, Dordrecht.
Rumelhart, D. and J. McClelland: 1986, 'On Learning the Past Tense of English Verbs', in J. McClelland, D. Rumelhart and The PDP Research Group (eds.).

ERIC REULAND

REFLECTIONS ON KNOWLEDGE AND LANGUAGE[1]

0. INTRODUCTION

One of man's most intriguing endeavours is that of acquiring an under-standing of the organizing principles of the human mind: the way the mind structures and represents knowledge about the world around us.

Many of the issues concerning these principles are not in the domain of ordinary science, but require new links among disciplines. Many fields are relevant to this enterprise, more than one could in practice bring together and integrate. At the conference from which this volume (and its two companion volumes) emerged, the following fields were represented: literary theory, philosophy, political theory, psychology of language and formal linguistics. Even though one could imagine other disciplines to be relevant as well, these may do. Establishing new connections among even a smaller number of fields may already be hard enough.

What I am going to say here will mainly reflect my personal perspec-tive, that of a theoretical linguist. But, I am convinced that at the present stage this is not necessarily a drawback.

1. CONCEPTUAL STRUCTURE AND MEANING STRUCTURE

The notion which plays a key role in bridging the gaps among the fields involved in the enterprise is *conceptual structure*. For a proper under-standing of this notion it is necessary to make a distinction between *concepts* and *meanings*.

In order to be able to deal with the world in which we live, we organize a mental model representing our knowledge of it. We split it up into objects, actions, situations, etc.; we ascribe properties, similarities, equiva-lences; we develop attitudes of fear or desire; we develop an orientation in space and time, and so on. From an evolutionary perspective, our con-ceptualization of the world should enable us to survive in it. This is the domain of concepts: concepts (in the present sense) belong to the model we form of the world in order to cope with it.

Meanings, on the other hand, are linguistic entities. The model they relate to is the domain of discourse, which we can view as a register providing us with variables taking concepts as possible values, and which imposes a rigid format on the information read into it. As a model of the world, the domain of discourse is impoverished to the extent that it makes possible fast calculations of complex relations in terms of the structure in

11

Eric Reuland and Werner Abraham (eds), Knowledge and Language, Volume I, From Orwell's Problem to Plato's Problem: 11—27.
© 1993 *by Kluwer Academic Publishers. Printed in the Netherlands.*

which they are represented. In brief, the domain of discourse is a model of the world for linguistic purposes, that is, for talking about it.

Let me illustrate the difference. Take words like **friend** or **brother** (in the sense of male sibling) and compare them with words like **man** or **chair**. **Friend** and **brother** are relational nouns. A friend or a brother is always a friend or a brother *of* someone. Words like **man** or **chair** are not relational. It can be shown that there are differences in the ways words of these types behave, and in the types of constructions in which they appear. For instance, a sentence like **Mary has every friend** is odd in a way in which **Mary has every book** or **Mary has many friends** are not; the same holds true for **there was a brother in the room** as compared to **there was a man in the room** or **there was a brother of Mary's in the room** (where the latter sentence explicitly expresses the relational character). An account of these differences is part of a theory of semantic structure.[2]

There are also completely different aspects to words like **friend**. Classifying a person, or a nation, as a friend or a foe may have immediate implications for the way you are going to treat him or her. There are things you do to foes, but not to friends. There may be cultures in which classifying someone as a foe implies a license to kill. In our culture, there does not appear to be such an implication on the person-to person level. On the nation-to-nation level, there is less evidence of any restraint. Anyway, the friend/foe classification represents a way to organize the world in order to cope with it.

In short, on the one hand, if I decide to refer to someone as my friend, this bears on the sentential patterns I can use; on the other, this has all sorts of implications as to how I will deal with the person. The former phenomena involve the meaning of **friend**, the latter the concept.

Prima facie, it does not seem likely that the two models and their organizational principles are identical; they are more likely to be associated with different mental faculties. There are specific reasons to think so. Grammatical gender is an important organizational principle in the domain of discourse (as discussed by for instance Bouchard (1984)). Yet, it is highly arbitrary with respect to sex (even in the domain of living creatures there is never a complete matching between gender and sex).[3] Here we have a linguistically pervasive (grammatical) category, which is utterly irrelevant as a category for coping with the world. The same can be observed about other grammatical categories. That the models we construct of the world for linguistic purposes contain categories which are irrelevant for coping with it, is what makes them so interesting for our purposes. Because such categories cannot be forced upon us by the world, they must reflect organizational principles of the mind itself.

Even if concepts and meanings belong to different domains, they must be linked. But how are they related? Clearly, not all meanings relate to concepts; although it is reasonable to say that functional elements such as the auxiliary **do** or the definite article **the** have a meaning, it would not do to say that in such cases there are corresponding concepts. For full lexical elements, like **friend, foe, chair, red**, or structures formed by standard compositional procedures the linguistic element does relate to a concept. The nature of this link is largely conventional.[4] In addition, one must acknowledge the existence of concepts for which there is no corresponding meaning: Those one can only identify by using non-conventional metaphors, puns, etc. This leads to the following point: there is a certain arbitrariness in the relation between *lexical items* (and the structures into which they combine), and *concepts*. Meanings are like pointers toward concepts. Sometimes the pointing is conventional, on other occasions we have to access all kinds of other resources in order to identify the concept being pointed at.[5] But of course, this is the beginning of a question rather than the beginning of an answer: for an answer, we need a precise theory of this pointing relation. It would already be an advance, if, at this stage, we could be sure that this is the right question to ask.

If conceptual structure exists alongside meaning structure, the question arises whether conceptual structure is governed by its own inherent laws, and actively organizes the world, or whether it is passive and, let us say, simply reflects the world. This is an empirical matter. Like any other subject matter, conceptual structure should be studied on the basis of its effects. Clearly, conceptual structure will make an important contribution to the way in which a sentence is ultimately understood. In cases where meanings are conventionally linked to concepts, it may be hard to sort out what contribution to the interpretation comes from the meaning structure, and what from conceptual structure. In the earlier discussion, it was implied that at least the heavily value laden aspects of the meanings (taken broadly) of elements like **friend** or **foe** belong to conceptual structure. Their relational character, however, is part of semantic structure. By itself, this does not decide where the boundary is. It is quite conceivable, for instance that certain of the distinctions discussed by Jackendoff are represented in meaning structure, whereas others belong to conceptual structure in our sense.[6] One can only seriously study conceptual structure by factoring out the effects and contributions of meaning structure to interpretation. Properties of lexical items that remain constant even if the concept indicated is varied, should be part of meaning structure (see Jackendoff's discussion of full versus light verbs[7]). Properties that are induced as soon as a certain lexical item points to some concept, and that emerge also if some different item points to the same concept most probably pertain to conceptual structure. This is why the study of metaphor may constitute such an important tool. Metaphor bypasses the conventional links between meanings and concepts. The associated concept must be identified on the basis of a variety of means, linguistic and non-linguistic, universal, or idiosyncratic. The ultimate understanding, however, requires that the identification is successful. Once a concept has been identified, it can be studied. Concepts thus found may provide a window into conceptual structure proper. Their properties may reflect the way in which we organize the world independently of grammatical structure.

Why is this issue important? Well, for one thing, metaphors seem to play a role in manipulation and persuasion. Speculating, it seems that pointing at a concept in a non-conventional way has the effect of delinking it, allowing it to be integrated into the structure anew, in another place of the network with different effects. The question how this process takes place is quite intriguing in itself. However, finding out how our mental models of the world are constructed and organized is also important for another reason already hinted at in the discussion of **friend** and **foe**. It may bear on our understanding of society.

2. ORWELL'S PROBLEM AND PLATO'S PROBLEM

Noam Chomsky has argued in a series of works that the selection and presentation of news by the media lead to a situation where crucial questions cannot be asked, since the background assumptions they require cannot be stated. The typically Orwellian case Chomsky cites is that of the American war in (attack on, aggression against) Vietnam. Even those doves who opposed the American involvement in Vietnam could not discuss the issue in these terms, because the concept of American aggression did, and to a large extent does, not exist for them: the combination is unthinkable, since anything America does is at least done with the best of intentions. Chomsky gives several other examples of successfully implemented Newspeak, e.g. what is referred to as the peace process in Latin America. It is important for Chomsky, that the unthinkability of an American aggression is not absolute; with access to the facts, and given sufficient explanation, anyone could become convinced that there was such an event. In our terms, under certain conditions of information, anyone would delink and reintegrate the concepts involved so as to make the combination possible. However, the media distribute the information in such a manner that for no significant part of the population will this happen. Relevant facts and opinions are either ignored or marginalized. Why is this so?

2.1. *Consensus and the Role of Information*

In his preface to *Knowledge of Language*[8] Chomsky contrasts Plato's and Orwell's problems. Plato's problem is to explain how we know so much, given that the evidence available to us is so sparse. Orwell's problem is to explain why we know and understand so little, even though the evidence available to us is so rich. Plato's problem is to be solved by "investigating the genetic endowment that serves to bridge the gap between experience and knowledge attained . . .". To solve Orwell's problem "we must discover the institutional and other factors that block insight and understanding in crucial areas of our lives and ask why they are effective". The character of the inquiry into these two problems is different, according to Chomsky. In the case of Plato's problem the questions ultimately belong to the sciences, with a pattern that is hard to discern, and requires abstract explanatory principles that can only be fruitfully investigated by specialists. In the case of Orwell's problem, the patterns are not difficult to discern, and "the explanation for what will be observed by those who can free themselves from the doctrines of faith is hardly profound, or difficult to discover or comprehend. The study of Orwell's problem, then, is primarily a matter of accumulating evidence and examples to illustrate what should be fairly obvious to a rational observer, even on superficial inspection, to establish the conclusion that power and privilege function much as any rational mind would expect, and to exhibit the mechanisms that operate to yield the results that we observe." Chomsky summarizes his position in the following words: "Plato's problem is deep and exciting; Orwell's problem, in contrast seems to me much less so."

These passages ignore the possibility that Orwell's problem has deeper roots than just the prevailing structure of society. In fact, Chomsky seems to have adopted an essentially empiricist approach that potentially obscures a proper understanding of Orwell's problem. The observation that the flow of information is often controlled and channelled by, and serves the interest of the existing power structure does not provide a sufficient basis to explain the nature of the prevailing structure by itself.

Chomsky's position here seems to be grounded in considerations of the following type. If one considers the language faculty, and other cognitive systems, one observes that the system is extremely well suited for carrying out tasks within a certain domain, whereas for tasks just outside that domain performance drops rapidly, or completely fails.[9] The knowledge normal humans have of the structural properties of their language is pretty much in the same range, regardless of differences in other capacities usually taken to constitute intelligence. Acquisition of this knowledge takes place within a set time span, and cannot be avoided under normal conditions. If acquisition has not taken place within that period, due to special conditions, it will be definitively hampered. Compare this to knowledge concerning the patterns of economic, political and social life, referring to it as 'social knowledge' (taking this term to cover the relevant parts of economy and politics as well). Anyone can study these phenomena on the basis of a conscious decision, acquire facts, find patterns, develop a position, etc. There is no indication that our possibilities to acquire knowledge in this field are restricted to a narrow specialized range. It is easy to imagine systems that are quite different from ours, which we might study with equal ease. The actual quality of the knowledge attained will heavily depend on the amount of attention the subject gives to his task, his endurance and perseverance, and his other intellectual qualities. There is no indication that there is any set period in which acquisition of this type of knowledge takes place. Hence, the differences with the process of language acquisition, or the process of learning how to use the visual system, seem enormous.

However, the parallel involved is not convincing. A number of questions and observations can be adduced. For instance, in the realm of cognitive capacities, a distinction must be observed between conscious knowledge and tacit knowledge. For social knowledge this entails that it is conceivable that there is a difference between what we can consciously know, and what we can actually put to use, just as in the case of Odd English. It is important to study the prevailing structure of society, and assess its influence on the flow and interpretation of information, but it is more important to ask the question how it came into existence, and how it is preserved. Why is it often so hard for people to resist policies with consequences recognized to be unacceptable? Why is it that people so often act against their own interest and even judgement?

Media play an important role in shaping public views. In the USA, as has been amply documented by Herman and Chomsky, the media are largely under the control of the dominating groups in society, specifically corporate business. As share holders they require maximum profit, hence no controversies are allowed which would cost revenues in the form of subscriptions or advertisements; as corporations, they expect no activities which could hurt their other interests. As a consequence, the press serves the interests of the elite groups by manufacturing consent in a way which is more effective than the coercion seen in totalitarian states.

The question is whether this is all there is to say. One may grant that in the USA the press presents a fairly uniform picture where the basic issues involving the structure of society and the world role of the USA are concerned, and that important aspects of the issues are never discussed. However, this cannot be the root of the matter. Like other social institutions, the press evolved. In earlier stages of society there was no such nation wide process of information moulding and distribution as there is today. But was there actually more diversity in the treatment of such issues in earlier times? If there wasn't there must have been an acting monopoly before there was a real one. If there was, the question arises why it disappeared.

In any case there must have other forces affecting the media. It is important to find out what these forces could be, how they influence the effect of the media, and in what direction. This requires considering cases where the media operate under different conditions than in the USA. Let us, therefore, briefly do so.

2.2. *A Comparison*

In comparison to the American press, the current Western-European press presents a picture of lively debate and dissent. Selection of issues, style of argumentation, are quite different from what is customary in the USA. Three questions arise: (i) Are there differences between Western Europe and the USA in the conditions under which the media operate? (ii) If so, is there a connection between these differences and diversity of opinions expressed? and (iii) Is there a corresponding difference in their role in or impact on society?

Here, I will limit discussion to the situation in the Netherlands. Chapter 1 of Herman and Chomsky (1988) provides a number of parameters by which one can compare. I should note, that my considerations in part rest on impressions; there is none of the thorough quantitative research Herman and Chomsky conducted, to back up my statements. Yet, by and large, the facts are well-known. This makes me confident that, after all, what I am saying is largely correct.

Although the situation appears to be changing, companies publishing newspapers are still largely independent. Unlike the situation in the USA, there is no standard connection of newspapers with large international companies. Although newspapers are dependent on subscribers and advertizing volume, there does not seem to be organized flak, as in the USA, where pressure groups try to systematically influence the course of the media by threats of boycott, etc. Where newspapers have merged to form one company there are charters protecting editorial independence, and not ineffectively so. Broadcasting companies are even less dependent on advertizers. Broadcasting time is allotted on the basis of membership, the costs are paid from membership dues, from a tax levied by the government on the possession of radio and TV receivers, and from advertizing. However, the broadcasting companies are not involved in the advertizing contracts. The potential advertizer buys time from a general body which then distributes the revenues over the broadcasting companies. So, the ties between advertizer and broadcasting company are indirect.

By themselves, these differences between the US and the Dutch situations are fairly striking. So the answer to question (i) is affirmative. To my mind, the answer to question (ii) is negative however. This is surprising in the light of what has been suggested, and calls for a reassessment of the role of the media.

The issue is nicely illustrated by the role of the Dutch media from the mid forties to the early sixties. In that period the properties of the Dutch media discussed above, were even more outspoken than nowadays. Yet, by the vast majority of the media the war against the Republic of Indonesia was simply represented as the justified repression of a rebellion. This was a fairly general perception; opposition was tantamount to treason. Similarly, the imminent constitutional crisis in the mid-fifties was not discussed in the Dutch press, despite the wide publicity it received in the neighbouring countries. For this whole period, one can say, that the basic purpose of the Dutch media was to represent, pressure and protect the Dutch existing order. This continued into the early sixties. Even then, in my recollection, the media were always lagging behind with respect to the cultural, intellectual and political developments.

This conformity cannot be explained in terms of the model of economic pressure as in the case of the USA. True, when the existing order was challenged by journalists with a more independent attitude, there were attempts from the side of the government to keep them in reign. However, from a present perspective, these attempts were so feeble, that it is surprising they could be effective.

One can only say that the source of the uniformity was a set of values and assumptions which were shared for rather different reasons, emerging from the way in which society was structured, and which by themselves where resistant to change.

These considerations lead to a reassessment of the role of the economic conditions. A different, and relatively independent economic base for the media does not necessarily lead to an increase in independence of opinion.

Consider next question (iii). To my mind, more recent developments (from the early sixties on) also require a reassessment of the impact of the press.

It is true that there are real differences between the media in the Netherlands and in the USA. This is substantiated by the facts reported in a survey conducted by Lex Rietman.[10] It can also be shown, however, that the effects of these differences are superficial at best.

During a considerable number of years, there has been a consensus within a significant proportion of the press with respect to a number of issues. These issues include acceptance of the use of soft drugs, legalization of abortion, opposition against apartheid and the Vietnam war, rejection of discrimination of e.g. guest workers, a liberal stand with respect to the admission of refugees. Among these issues a number of types may be distinguished: rights one might want to exert oneself (the right to smoke marihuana, the right to undergo an abortion), opinions that won't hurt (protesting apartheid in a far-away country, or opposing the war in Vietnam, where no particular Dutch interest was at stake), and liberties one has to grant others (varying from housing for guest workers or refugees, to granting them the right to their own style of living). A broad-minded view on the issues of the first two types has been fairly generally adopted. However, with respect to the last issue, the flow of information has not prevented racial tension, riots, protests when housing was allotted, etc. In fact, although racial discrimination has been condemned by the press for at least the last three decades, this appears to have had a surprisingly limited impact. The ways in which protests against housing of refugees are currently being treated and commented on even gives rise to the impression that the press is reassessing its position under popular pressure.

This indicates that the impact of the debates in the media is superficial, and does not really affect the underlying values of a given society. With respect to many issues there appears to be consensus, even where it cannot have been manufactured by implicit rules of what is to be expressed in the press. In fact, the Dutch situation should make one consider the possibility that there are aspects to the structure of society that are immune to the type of influence information may exert. These aspects may well be more constitutive of human society than one would generally be ready to admit.

These considerations lead away from the idea that the media play a decisive role in shaping the public opinion. They indicate that popular consensus is the source, rather than the result of press consensus.[11] The media reflect the values of the society they are part of. From this perspective the apparent liveliness and diversity of the debates in the Western-European media may well be little more than a reflex of the role which the intellectual debate plays in Western-European culture in general, prominent, but rather 'unverbindlich'.

3. ORWELL'S PROBLEM AND COGNITIVE STRUCTURE

To the idea that certain aspects of society are immune to change under information, a more general consideration may be added: There is a disquieting parallel between the constancy of the structure of language during the periods of human development we have evidence of, and the constancy of the structures of the regimes humanity has gone through. Whatever the regime, democracy has never extended beyond the group that was in a sense 'part of the regime'. In this respect the democracy in the USA or the Netherlands is neither better, nor worse, than, let us say, the democracy in Athens. At best the size of the group that takes part is larger, as is the proportion of the population it represents.

It is evident that in coping with their physical environment people make use of innate cognitive structures.[12] There is no reason to exclude the possibility that in coping with their human environment people also make use of organizational schemata that are part of their cognitive structure. It is very clear, that in coping with small scale situations, there may be vast discrepancies between the conscious knowledge people may have about relevant factors and options available, and what they are actually able to do and put to use. That is, they may act 'irrationally', know they act 'irrationally', and cannot but act 'irrationally'. There is no reason to expect that the same type of irrationality does not characterize their actual behaviour in the large scale matters of society. Notice, that the issues raised here do not necessarily lead to more intractable questions concerning human motivation (Descartes's problem, as Chomsky called it).[13] It is rather the limits posed by human nature that concern us here. If the actual behaviour in such matters takes place within the confines set by innate schemata it is of the utmost importance to determine what these schemata are, and how they work.

Assuming that there are such schemata, their effect cannot be uniform. There must be variation in the way they are filled in. Consider a simple 'us'/'them' schema, which prima facie seems to be a component of human categorizing (and which may be actually made use of by those influencing opinions). Fundamental to any society appears to be the distinction between those that do, and those that do not belong to it, and within it the distinction between those that are part of the power structure and those that are not. It looks as if some schema based on the 'us/them' distinction is constitutive of human societies in general. Hence, one should seriously consider the idea that it is part of the basic cognitive endowment in terms of which we cope with the world. Notice, that for instance the political effects of the various ways this schema can be filled in will depend on the size of the 'us'-group, and the way it coincides with the 'us'-group of others. This schema will of course be most effective, if, for large numbers of individuals, it selects a group sharing political institutions and means of power. Note, that this view does not imply that societies cannot change. People differ as to the extensions of their 'us'-group, and the extensions themselves may change.

But, suppose that the relevant feature also categorizes properties of 'us' and 'them', so that every property of 'them' will be 'typically them' and every property of 'us' 'typically us'. It is clear that in such a case no amount of information will be able to change the picture. The picture will only change if somehow 'them' or part of 'them' can be reclassified as 'us'. If this process is to be influenced, one should know in the first place what factors determine application of this schema. The same holds true for other schemata in human categorizing.

It seems to me that a substantial part of what seems to be Orwell's problem resides in the well-known tendency of humans to interpret new information in terms of what they already know, and, by implication, as supporting any prejudices they have. The more threatening information is to what is felt to be the essence of one's living conditions, the more readily it will be dismissed, and in fact be held against the one who provides it. This, in turn, reflects on the way the information is conveyed, if at all. For instance, it is safe to assume that many journalists (and politicians) realize that a reasonable distribution of wealth among the world's population is impossible unless the western world gives up many of its economic privileges. Yet, serious discussion of this major issue is rare. Where discussion is conducted it seems to be kept within the confines of the intellectual debate, where it safely stays at an abstract level. The issue is avoided, not because one is unaware of the facts, but because the message, if effective, may put the messenger at risk.

There must be more to Orwell's problem, though. A distinction is necessary between general conceptual schemata and principles governing social interaction, although their effects may be produced jointly. Universally, people appear to form societies with rules of varying complexity. But note that, although the complexity may vary, it is easy to imagine forms of society that never occur.

In general, just as the observable differences among languages should not obscure the fact that they are based on a restricted set of underlying principles that are rooted in the mind, the observable differences among human societies and cultures should not keep one from entertaining the hypothesis that in the domain of human societies and cultures shared principles are also involved that belong to our common cognitive endowment. As such they should be the subject of empirical investigation on the same footing as the principles underlying our capacities in other areas.

The purpose of such investigations is clear; unless we come to understand the unconscious principles underlying the way we categorize and organize the world, we will never be able to effectively change it. In terms of such a project, Plato's problem and Orwell's problem may well turn out to be different sides of the same coin.

4. WHERE DO ORWELL'S AND PLATO'S PROBLEMS TOUCH?

The central area where the Plato's and Orwell's problems touch is given by the question of why people fail to acquire certain types of knowledge.[14] Just as human nature may provide a clue to why certain learning takes are so easy, it may also provide a clue to why certain learning tasks are so hard.

Notice, that even within the domain of knowledge of language there are Orwellian aspects to Plato's problem. One case in point is the fact that enriching the language data does not appear to speed up the acquisition process. So, instead of stressing the poverty of the data, one might as well stress the fact that the process of language acquisition is to a considerable extent insensitive to the data (of course not completely, otherwise all languages would be identical).[15]

Another point is that much of the discussion of language acquisition assumes that there are limits to the class of learnable languages. Learnable,

that is learnable under natural conditions of exposure (perhaps 'passive
learning' is a useful phrase, which may be contrasted with 'active learning'). This is an empirical question. Suppose one were to devise a form/
meaning mapping outside the class of natural language. The empirical
claim is that if children were exposed to data from such a language, they
would only partially acquire it, or not at all. Notice, that the thought
experiment requires that this language can be learned and even handled
(by the people who were to expose the children to these data). But this is
irrelevant for the logical status of the problem, because people do learn
how to handle nonnatural form/meaning mappings (logical languages,
programming languages; see Chomsky (1965)[16] where it is pointed out
that systems that cannot be acquired under 'natural' conditions may still be
acquired if approached as part of a game, or an intellectual puzzle).[17]
Under the conditions sketched and given what we know about language
acquisition there is no reason to expect that 'enriching the data' short of
explicit instruction, would be of any help.

   Let us go back now to Orwell's problem and the domain that is its
paradigm instantiation, social knowledge. The assumption underlying the
idea that Orwell's problem is intellectually unexciting, is that there is no
qualitative difference between social knowledge acquired on the basis of
active learning, and that acquired on the basis of passive learning triggered
under exposure. What I want to bring out is the fact that this is an empirical assumption. At least two empirical issues are involved. First, is social
knowledge in fact acquired both by passive and by active learning?
Second, if it is, are there qualitative differences?

   As to the first issue, without doubt children acquire considerable
knowledge of their social environment with no explicit instruction. And on
the basis of it they are very capable of efficiently handling their immediate
environment. This type of social knowledge may be similar in status to
what Fodor called our surprisingly effective 'grandmother psychology', but
perhaps not as effective.[18] As far as I can see, the real issue is not so much
whether some knowledge is acquired in that manner, but how far that
mode of acquisition extends. The second issue is open, I think, and should
be investigated. But the possibility of a positive answer should be seriously
considered.

   In investigating this issue, one should keep in mind that different modes
of acquisition may differ not only by the conditions under which they
operate, but also by the way in which they project data. If D is some set of
data and D_X the knowledge structure to which D is projected by some
learning device X, it is conceivable that D_{pass} differs from D_{act}. One of the
dilemmas facing us is why some people do appear to acquire relevant
social knowledge, and others don't (note that this is independent of the
question how they act on that knowledge). It is not necessarily a matter of
general intelligence. It has been noted by Chomsky that certain people on

the one hand acquire highly intricate and abstract knowledge about sports games, and the structure of sports competitions, while they are lacking in relevant knowledge of their socio-political situation, "as if it were a conscious decision". From the present perspective this might come out as a difference in the learning strategy employed. That is, people may take certain domains to be not amenable to active learning. They do not apply that strategy, hence do not acquire D_{act}.

Let this suffice for the moment. I am not under the impression that at this stage I could say anything substantive about the issues involved. These are all matters for empirical research. The point is that they are important and as rewarding as the investigation of the question how knowledge is acquired in any field. My contribution at this point just boils down to the suggestion that a rationalist approach may be relevant not only for Plato's problem, but for Orwell's problem as well.[19]

Following the lines set out above, we are led to the concept of what one might call a 'natural' society, similar in status to that of a natural language. Accepting this term, in spite of its unwanted historical connotations, the question comes up what the status of such a concept might be.

We know that learning language is something human beings are good at. But, this does not necessarily hold true for systems derivative of natural language, varying from 'specialist' languages, such as legal language, to formal languages. In the case of formal languages it is fairly obvious that they could not be acquired under mere exposure. With respect to specialist languages this may vary. In principle one may assume that they stay within the confines of natural language. It is quite conceivable, however, that some cross the border and end up containing non-natural subparts. One may expect this to result in a situation where many people fail to cope adequately with such a language. And indeed, many people leave specialist languages to the specialist. To the extent that 'the language' (here a sociological concept) of some society contains that specialist language as a subpart, people will in fact surrender their say in that domain to the specialists involved.

The same may happen to societies. The upshot of applying rationalist ideas to the domain of Orwell's problem is that it opens the possibility of distinguishing between rules of society we are good at acquiring and handling, and rules of society we are bad at. Just like the *language* of a society may develop non-natural subparts, it is conceivable that a society itself develops non-natural subsystems; non-natural in that they are based on rules and principles we are bad at handling. As a consequence, control will be transferred to those specialists who are capable of handling the system. Clearly, this is what is characteristic of bureaucratic societies. There is an inherent danger in such systems. It seems obvious that a society can only function well, if its participants are able to handle its rules. If a specialist ruled society is put under pressure it will be in danger of breaking up, since the specialists may not be able to come up with the type of solutions the other participants can handle. Therefore, suppression and violence may seem to provide the only possible avenue for those in power. The conclusion should be, that however complex and overwhelming the problems of our societies are, one should be wary of solutions that are beyond our natural capacities to handle.

5. FREEDOM OF CHOICE

If the optimal structure of society is not independent of human nature, the question comes up whether this infringes on the idea of free will. This, however, is not the case. Free will belongs to the domain of human action or motivation, and we are not discussing a theory of that domain.

Just as the existence of innate principles of universal grammar does not violate one's freedom to say **X** or not to say **X** (nor the freedom to construct ungrammatical examples), the existence of innate principles of a universal 'social grammar' does not violate the freedom to construct societies violating them. Only, there is a cost if this freedom is exercised. It seems to me that history provides evidence for this.

Free will essentially presupposes control. It cannot be a matter of free will whether one fully acquires some system, be it a language system or a society system, under natural conditions of exposure. Free will enters into decisions: decisions to attempt, and decisions how hard one will attempt. So it does enter into the decision whether one will acquire some system as an intellectual task. (Even so, a system may be too hard to acquire. For instance, if one decides to have a go at mathematics, despite being bad at it.) Similarly, it is a matter of free will to decide to unravel the workings of society or decide to try and change it.

In general, freedom of will as applied to actions seems to involve (at least) two components: 1) the freedom of the decision to try to perform the action; and 2) the control one has over performing the action. Whereas in the cases considered the contribution of each component to the action is clear, it is conceivable that in some cases the distinction becomes blurred. Lack of control may make it senseless to speak of freedom to decide to try, complete control may make the freedom to decide to try as good as the freedom to do. It is cases of partial control that are the interesting ones. It is to be hoped that investigation of these cases will provide us with insight in the workings of the human mind.

6. CONCLUSION

Starting out from an elucidation of the notions *meaning* and *concept* we reviewed a number of considerations bearing on the relation between Plato's and Orwell's problems as sketched in Chomsky (1986). We argued that the mechanisms underlying the control of information, but also its impact, may well be more directly related to human nature than was originally suggested.[20] On the one hand, shared values may be as effective as owning shares as a means of controlling information. On the other, popular consensus may be the source, rather than the result of press consensus. In short, in the cases discussed the media reflect the values of the society they are part of.

In order to understand the interaction between values and information it is essential to investigate how man acquires knowledge about his social environment and what are the constraints on his capacity to deal with information of this type. In this respect 'social competence', need not be different from linguistic competence, or from our competence in performing complex mathematic calculations. If Orwell's problem is understood as the problem why (in the domain of social knowledge) we know and understand so little, even though the evidence available to us is so rich, its source may well be that people are bad at handling certain types of data, or need special triggers in order to apply the necessary strategies. But then it is essential to find out what the principles underlying these restrictions are.

If the picture sketched here is true, it is fair to conclude that the study of the nature of Orwell's problem may be as intellectually rewarding as the study of language, and more crucial to our survival.

NOTES

[1] I am very grateful to Harry Bracken, Noam Chomsky, Jan Koster and Ken Wexler for stimulating discussion and to Werner Abraham, Harry Bracken and Jan Koster for their careful reading of this manuscript. I would also like to thank an anonymous reviewer for his encouraging comments. Needless to say that they do not necessarily agree with any of the ideas expressed in this article.

[2] Perhaps it is not impossible to imagine contexts in which such sentences could be used. However, such use is always heavily dependent on specific features of the preceding context. Similar remarks apply to measure expressions, cf. the contrast between **this road is many miles long**, versus ***this road is every mile long**. For discussion of the special quantificational properties of relational nouns see De Jong (1987) and Keenan (1987).

[3] The same holds true of the animate-inanimate distinction in languages where it has grammatical reflexes.

[4] This is not to claim that the relational character inherent in the meaning structure of the lexical item **friend** cannot be linked to any property of the concept it is associated with. However, as in the case of sex versus gender, one will expect that the match is not complete, i.e. lexical items should occur that are conceptually relational, but not linguistically so, and vice versa. Note furthermore, that much of what one could say about a concept will never reflect on the grammatical properties of the associated lexical item. Inspecting a concept by itself will not give any clue as to what is linguistically relevant. Alternatively, nothing in the meaning of **foe** will tell us whether foes may be robbed, killed, etc., or not.

[5] This entails that meaning structure is more than just thematic structure. The meaning structure of a lexical item must contain some information enabling it to successfully point at a concept. However, this extra may be just a conventional index.

[6] Cf. Jackendoff (1983) and Jackendoff (1987a).

[7] Jackendoff (1987b).

[8] Chomsky (1986).

[9] Consider a language Odd English that deviates from English in some mathematically trivial property, e.g. that the length of its sentences, measured in words, must always be an odd number, with a grammatical dummy that can be used to guarantee that with any number of content words always a sentence can be formed that has the property. It is clear that the language faculty would not enable humans to perform better than chance on this requirement.

[10] Rietman (1989).

[11] From this perspective also the limited effect of the press in totalitarian states may be more readily understood. In so far as the propaganda did not reflect the values of the societies it was directed at, its effect was negligeable. As we now know, the effect of the propaganda in the USSR and the countries it controlled was even less than people assumed it was at the time the system was still in effect.

[12] See Chomsky (1988) for discussion.

[13] Cf. Chomsky (1988).

[14] This question must be distinguished from the question why people may fail to act on knowledge they have. The latter question involves human motivation. As noted above, questions of motivation per se (Descartes's problem) may well be intractable; hence, they should not concern us here. There is an important borderline case, though, namely when people cannot act on knowledge they have. This will be briefly discussed below.

[15] Cf. Chomsky (1988).

[16] Chomsky (1965).

[17] There is an interesting empirical angle to the experiment. It presupposes a form/meaning mapping that is rich enough to talk about the world, and that can be use 'naturally' (more or less as a second language) in front of the children after having been acquired unnaturally (by the people that devised it). It is conceivable that no such system exists. That is, that a system can be acquired naturally if (and only if) it can be used naturally. This cannot be taken for granted, though. However, for the main argument we need not go into these issues.

[18] Fodor (1983).

[19] The question might be raised whether a rationalist approach to the limitations on the acquisition of social knowledge does not lead to an elitist view of society, with all the dangers involved. However, the present view does not lead there, or to anything of that kind, for the same reasons that rationalism in general does not lead there, as extensively discussed in Bracken (1984). The essence of rationalism is that it takes the human being seriously as it is; accepting that human beings qua human beings are good at performing certain tasks and bad at performing others. If there is a political implication it is that human beings had better form their societies taking into account what is known about their capacities and their limitations.

[20] Still there is no reason to doubt that the flow of information in a society often serves the interests of the existing power structure.

REFERENCES

Bouchard, Denis: 1984, *On the Content of Empty Categories*, Foris, Dordrecht.

Bracken, Harry: 1984, *Mind and Language*, Foris, Dordrecht.

Chomsky, Noam: 1965, *Aspects of the Theory of Syntax*, MIT Press, Cambridge, Massachusetts.

Chomsky, Noam: 1986, *Knowledge of Language: Its Nature, Origin and Use*, Praeger, New York.

Chomsky, Noam: 1988, *Generative Grammar: Its Basis, Development and Prospects*, Kyoto University.
Fodor, Jerrold: 1983, *The Modularity of Mind*, MIT Press, Cambridge, Massachusetts.
Herman, Edward S. and Noam Chomsky: 1989, *Manufacturing Consent: The Political Economy of the Mass Media*, Pantheon Books, New York.
Jackendoff, Ray: 1983, *Semantics and Cognition*, MIT Press, Cambridge, Massachusetts.
Jackendoff, Ray: 1987a, *Consciousness and the Computational Mind*, MIT Press, Cambridge, Massachusetts.
Jackendoff, Ray: 1987b, 'The Status of Thematic Relations in Linguistic Theory', *Linguistic Inquiry* **18**(3), 369—411.
De Jong, Franciska: 1987, 'The Compositional Nature of (In)definiteness', in E. J. Reuland and A.G.B. ter Meulen (eds.).
Keenan, Edward: 1987, 'A Semantic Definition of "Indefinite NP"', in E. J. Reuland and A.G.B. ter Meulen (eds.).
Reuland, Eric J. and Alice G. B. ter Meulen (eds.): 1987, *The Representation of (In)definiteness*, MIT Press, Cambridge, Massachusetts.
Rietman, Lex: 1989, *Over objectiviteit, betonrot en de pijlers van de democratie: De Westeuropese pers en het nieuws over Midden-Amerika*, MA thesis, Instituut voor Massa-communicatie, University of Nijmegen, Nijmegen.

NOAM CHOMSKY

MENTAL CONSTRUCTIONS AND SOCIAL REALITY

The issues proposed for discussion today are very broad, reaching well beyond what I could hope to address. I will keep to only a few of them, but first, perhaps it would be useful to try to situate these within the terrain sketched out in the preliminary statement outlining today's proceedings.

Let us understand *Plato's Problem* to be the question of how we come to know and understand so much, given that we have so little evidence; and *Orwell's Problem* to be the question of why we know and understand so little, given that the evidence before us is so rich. There is no contradiction, since the problems arise in different domains, Orwell's being restricted to societies and their workings; in the most interesting case, to features of our own society.

There are, to be sure, many other domains where we have a great deal of evidence but understand very little: a classical example is the matter of choice of action. Elsewhere, I have suggested that we distinguish between "problems," which fall within the range of our understanding, and "mysteries," which do not. The distinction need not be clear-cut, but it is nonetheless quite real, at least if humans are part of the biological world, not gods or angels. In the case of other organisms, we have no difficulty in recognizing that their cognitive capacities have scope and limits, both deriving from the same biological endowment. And we take for granted that the same is true of the growth of organisms, including humans "below the neck," metaphorically speaking; the same biological endowment that guides the human embryo to become a person prevents it from becoming an insect or a bird, under some modification of the nutritional environment. There is no reason to doubt that the same is true of the capacities of the mind/brain to construct systems of understanding, evaluation, belief, knowledge, and interpretation.

Among the cognitive structures that the mind attains we find some that grow effortlessly and without awareness, and others that are developed laboriously and through conscious effort, at least in part. Human language is perhaps the best understood case in the former category, the physical sciences the prime example of the second. In both categories, the cognitive capacities are rich enough to provide many striking cases of Plato's problem; and as a consequence of the same initial endowment, they leave us "without intelligence enough," in Descartes's phrase, to approach other questions, some of which may be quite important to our lives.

I will assume that the distinction between problems and mysteries is

29

Eric Reuland and Werner Abraham (eds), Knowledge and Language, Volume I, From Orwell's Problem to Plato's Problem: 29—58.

real, and species-relative: what is a mystery to a rat or pigeon may be a problem for us, and conversely. I will assume further that the distinction just indicated within the domain of problems is also real. Within this domain, then, we may distinguish such "natural" tasks as acquiring a language or a grasp of the properties of objects located in three-dimensional space, and others that fall, to a greater or lesser degree, within the range of what we might call our "science-forming capacities." Here the distinction may turn out to be quite clear-cut, as a matter of biological fact, if indeed the mind/brain has the highly modular structure that current understanding suggests. Let's assume this to be so, at least as a good first approximation. I will thus distinguish between "special purpose" systems such as the language faculty and what we might call "the rational faculty," referring by that term to the elements of the mind/brain that provide what Charles Sanders Peirce called our instinctive capacity of "abduction" and evaluation, of forming intelligible theories and subjecting them to test, something that we can occasionally do.

Notice that this is not quite Jerry Fodor's conception of modularity; I do not take the language faculty to be an input system (though there are input and output systems associated with it), and I assume that the rational faculty (a — or the — central system in his terms) has its own special design. Note further that modularity assumptions are controversial. It is widely alleged that language has no special structure; its properties are determined by undifferentiated mechanisms of general intelligence. Thus, standard arguments concerning alleged social and indexical elements in meaning commonly draw conclusions about natural language from what is taken to be rational procedure for choice of language in the natural sciences, a step that is legitimate only if modularity is denied; Donald Davidson goes so far as to hold that there is no use for "the concept of language," for "shared grammar or rules," for a "portable interpreting machine set to grind out the meaning of an arbitrary utterance," and we are led to "abandon . . . not only the ordinary notion of a language, but we have erased the boundary between knowing a language and knowing our way around in the world generally." For reasons discussed elsewhere, the arguments presented seem to me faulty and the conclusions untenable, and I will keep to the well-supported position that the language faculty does involve special design.[1]

In these terms, we can return to Plato's problem and Orwell's problem. Both arise in domains that fall within the potential of the mind/brain, given its specific structure; sometimes a matter of yes-or-no, sometimes of more-or-less, as a matter of biological fact. But their status is quite different. In one case we seek to explain the success in achieving rich and complex cognitive states; in the other, to understand the failure to deal with far simpler issues, which would readily be mastered were it not for interfering factors. In his best-known work, Orwell dealt with a special case of the problem: the measures of control in a totalitarian society, those that prevent people from seeing, or at least saying, that $2 + 2 = 4$, to take Winston Smith's example — though we might add that freedom means the right to say that $2 + 2 = 5$, and to shout it from the rooftops, as Orwell doubtless would have agreed. There is another variant of the problem that is intellectually more interesting, and far more important for us: namely, the mechanisms of thought control in relatively free and democratic societies. Had Orwell written a novel about this topic, it would not have been appreciated, and he would have faced silent dismissal or obloquy instead of acclaim.

The latter variant of Orwell's problem is the more interesting one because the mechanisms of indoctrination and control are less transparent in free and democratic societies than in those where orders are issued from the Ministry of Truth and Big Brother wields a very visible knout. In fact, to the extent that a society fits the totalitarian model, it need not be overly concerned with control of belief and thought. The leadership can be "behaviorist"; it doesn't matter much what people think, as long as they can be trusted to obey. Insofar as the state lacks power to coerce, control of thought becomes a higher priority. And the measures adopted are, correspondingly, more interesting to unravel. Some close observers of the Soviet Union hold that its propaganda system is moving towards the more efficient western model as controls are relaxed, with a "plurality of 'responsible' viewpoints" kept within the range of the "authorized pluralism" instead of "incessant trumpeting" of transparent propaganda; then "power, which was the overt source in the old-style propaganda, under the new style becomes rather invisible," "the functions of ruling the country and informing the public become then dissociated in the popular mind," and some variability of messages "can be allowed on the plausible presumption that most of them will be forgotten" and overwhelmed by the desired doctrines, as in the more sophisticated western model (Jedlicki 1989).

Nevertheless, I am not convinced that there is much intellectual depth to Orwell's problem even in the more interesting and complex case of relatively free societies. Whether this belief is right or wrong, it does not imply or suggest that "the mind is minimally structured" in the domain of social reality, as suggested in the opening statement. Independently of Orwell's problem, it seems reasonable to suppose that the human mind is richly structured in the domain of social reality, that some faculty of the mind/brain develops a conception of the social world that enables us to function within it. If so, we have yet another case of Plato's problem. Furthermore, the belief that Orwell's problem is not very difficult to solve does not entail that "the ways in which power and privilege function can be explained in terms of a silent control of information channels" (quoting again the preliminary statement). Rather, control of information channels is one device for ensuring that power and privilege will be able to function without interference from the annoying public, a lively concern among educated elites in Western societies for centuries. I am also skeptical about some of the comparative judgments expressed in the introductory comments, for reasons to which I will briefly return.

I hope these remarks help clarify my own standpoint. I will proceed on the assumption that humans are part of the biological world, that the mind/brain operates in the manner of other biological systems. There is a rich initial endowment and, it appears, a highly modular structure, as in the case of any complex system that we know. No evidence has been presented to suggest that there exist "generalized learning mechanisms" that apply over all domains. It also seems unlikely that such notions as "conditioning" have much of a role to play in the inquiry into how cognitive systems develop — even if the phenomenon exists as something other than an experimental artifact. Dogmatic assertions about the matter in the literature are rather surprising, particularly on the part of those who call for a "naturalistic" stance and take the sciences to be "first philosophy." It is, furthermore, an open question to what extent anything resembling "learning" as ordinarily understood can be detected in cognitive growth, and the idea that some kind of learning mechanisms, if they even exist, will account for such disparate tasks as acquiring a language, developing the ability to interpret objects in motion, and discovering (or learning) quantum mechanics seems unlikely, in the light of what is at all understood today. Speculation to the contrary is widespread in recent literature in several fields, but the conceptual and empirical difficulties that at once arise seem rather serious, to say the least.

To the extent that we come to understand the nature of specific faculties of the mind, we will be able to determine what constitute problems and mysteries for them. In the case of the language faculty, we know enough to be able to design language-like systems which, we predict, the mechanisms of the language faculty will not be able to acquire, because they will always make "the wrong guesses." If we understood more about the rational faculty, we should be able to do the same. We should be able to design possible worlds — which might include elements of the actual world — that pose tasks that the human rational faculty will always fail to solve; abduction will always fail to yield intelligible theories that come close to dealing with the empirical materials at hand, suggesting that we are entering the domain of mysteries, not humanly accessible problems. We might even learn why the bounds fall where they do, just as we can learn why some language-like systems fall beyond the scope of the language faculty. Failure of abduction is a common feature of inquiry, in fact, the normal case. It may show no more than that the questions have been wrongly posed or lie beyond current understanding, or that we are approaching mysteries for reasons that might be established by rational inquiry.

There is no contradiction in the belief that our rational faculty may suffice to provide insight into its scope and limits. Thus inquiry might reveal that the theoretical structures provided by the rational faculty have certain properties, as a matter of biological necessity, and that some elements of the world do not conform to these conditions. An organism capable of constructing only the rational numbers would find that its physics and mathematics have odd gaps, and might even characterize its own capacities (say, algebraically), without ever being able to grasp how these gaps might be filled. In our case, we might discover that the human abductive capacities permit us to deal with input-output systems, deterministic systems and those that have random or probabilistic elements, but nothing more. And we might discover that the Cartesians were correct in their belief that ordinary human behavior falls beyond these limits, that unlike automata, humans are only "incited or inclined," not "compelled" to act in certain ways when their internal arrangement and the environment are fixed. They might also prove correct in the belief that the best indication that another organism has a mind like ours lies in what we may call "the creative aspect of language use," the ability to produce utterances that are new and undetermined by stimuli and internal configurations (even probabilistically), that are appropriate to situations but not caused by them, that are coherent and evoke in us thoughts that we might have expressed the same way. And it may be, as Descartes sometimes suggested, that the ability to understand all of this lies beyond the grasp of our rational faculty — or in his terms, of mind itself.

A language-like system that falls beyond the scope of the language faculty might be unravelled by the rational faculty, much as some properties of the physical world have been discovered over centuries of effort. That other systems will fall beyond the scope of the rational faculty as well is inevitable, assuming, again, that humans are part of the biological world. Our natural sciences might be thought of as a kind of chance convergence between the cognitive structures of the mind and properties of the world, and neither biology nor any other domain of knowledge provides us with any strong reason to suppose that this convergence will include many questions that might interest us.

A word may be in order about contemporary skeptical arguments questioning realist assumptions and denying that the idea of convergence to truth is even a meaningful goal for science. On one such view, we create versions of many worlds by the use of symbols, physics being one version, cubist painting another; we thus approach Montaigne's view of science as just "sophisticated poetry." It is plausible enough to hold that "worlds are as much made as found" (Nelson Goodman), if by that we mean that our

cognitive faculties play a role in determining the content of experience and
what we construct to depict or interpret or explain it. But I see no serious
reason to question what Richard Popkin calls the "constructive skepticism"
of Gassendi and Mersenne in their reaction to the skeptical crisis of the
16th—17th century, recognizing that "the secrets of nature, of things-in-
themselves, are forever hidden from us" and that "absolutely certain
grounds could not be given for our knowledge" though we do "possess
standards for evaluating the reliability and applicability of what we have
found out about the world" — essentially the standpoint of the working
scientist. Proceeding on these assumptions, we will inquire into the cognitive faculties themselves, regarding them as just another part of the natural
world that we hope to understand, and also trying to determine just how
they contribute to the construction of experience and formation of explanatory theories, including the ones we are attempting to construct. The
lack of indubitable foundations need not lead us to reject the working
assumption that there is an objective reality to be discovered, of which we
have at best a partial grasp.[2]

To clarify further, I have been speaking freely about the human mind,
but with no dubious metaphysical assumptions beyond those of normal
science. The brain, like other complex physical systems, can be studied at
various levels of abstraction from mechanisms. This study can proceed
without knowledge of what the mechanisms are, and may even be a guide
to discovering them, much as 19th century chemistry, with its abstract
notions of valence, halogens, benzene rings, and so on, provided a guide
to the physicists who had to revolutionize their understanding of the
physical world to account for the facts and principles discovered at the
more abstract level of inquiry. Some propose that such abstract entities as
neural nets will yield insight into the functioning of the brain; in some
domains, there is recent work suggesting that they may be right. Other
work indicates that computational-representational systems allow us to
capture and explain many of the phenomena of human thought and action.
For the latter work, mentalistic terminology is often used. There is no
harm in this, as long as we understand that this is simply normal science.

It is difficult to see how the mind/body problem can be given an
intelligible formulation in anything like its classical form. The Cartesians
could formulate the problem, because they had a reasonably clear conception of body, of the material world; namely, in terms of an intuitive
"contact mechanics." This conception did not survive Newton, and in the
subsequent era, there is no notion of body apart from what we discover
about the world, at whatever level of representation we adopt. Correspondingly, the question of what lies beyond the domain of body does not
arise, there being no fixed concept of the material world.

Suppose it turns out that such abstract systems as neural nets exhibit
properties of the computational-representational systems constructed in

the study of the language faculty, a common claim in the connectionist literature. Will this show that such entities as words, empty categories, morphological rules, and so on, do not exist in the material world (there being no other world)? Only if we also believe that the discoveries of modern physics show that there are no elements, molecules, liquids and solids, proteins, rivers, galaxies, and so on, and no principles expressed in terms of such entities. We naturally seek to find links among the various levels at which we study properties of the world. No one seriously doubts that the computational-representational systems of mind are realized in mechanisms of the brain. The discovery of such mechanisms may help us understand why the metalistic theories are true, insofar as they are, by establishing the links we seek. It is possible that connectionist abstractions capture enough of the properties of the neural mechanisms to help bridge the gap between computational-representational theories and whatever the brain mechanisms turn out to be, thus serving, as some have speculated, as a kind of "implementation system." It is always conceivable that connectionist models might prove the computational-representational systems to be erroneous — that is, based on assumptions that are wrong or unnecessary to explain the phenomena at any level. That possibility, for the moment, is hardly worth considering, given the empirical and conceptual failures of existing approaches and the question-begging character of such partial successes as there are.[3] For the time being, it remains true that in the domain of language at least, apart from computational-representational theories, there are no approaches with even limited plausibility, and the success of these theories is not inconsiderable.

Plato's problem arises when we discover that organisms attain a complex and highly articulated state in a relatively uniform manner. The transition to this state depends on the interaction of two factors: the external environment and the internal structure. The problem is to identify their respective contributions. In the interesting cases, the course of growth and development is significantly underdetermined by the external environment so that rich and specific structure must be attributed to the organism to account for the character of the transition. Thus one assumes, even without any understanding of what is involved, that the development of the human organism from embryo to adult is guided largely by an internal program that is specified in the initial biological endowment; the environment provides no information to determine this specific course of development. Suppose someone were to suggest, for example, that the embryo develops arms rather than wings because of the character of the nutritional input, or that children undergo puberty at a certain age because they are reinforced for doing so — through peer pressure, perhaps. The proposal would not be taken seriously, not because the mechanisms are understood, but because the problem of "poverty of the stimulus" is so severe as to leave no serious possibility beyond internal guidance under the triggering and marginally shaping effect of the environment. In the case of cognitive growth — that is, the growth of components of the mind/brain — the logic is similar, and similar conclusions seem inescapable.

In the case of language, there is overwhelming evidence that children are not initially adapted to acquire one rather than another language. That is, the language faculty of the mind/brain has a complex *initial state*, genetically determined, which is close to uniform across the species, severe pathology aside, and is apparently unique to the human species in essentials. Quite meager environmental inputs lead to the growth of a highly articulated *steady state*, after which changes appear marginal, basically acquisition of new lexical items in an already determined format. Many features of language structure have been studied which appear with virtually no relevant evidence, or none at all. There is good reason to believe that at the steady state, the language faculty incorporates a *generative procedure* that assigns *structural descriptions* to expressions, where each structural description expresses the properties that determine the form and meaning of this expression, insofar as these are determined by the language faculty. The only plausible assumption is that the basic character of these generative procedures is fixed in the initial state, with only limited variation possible.

The generative procedure that constitutes a language, understood as a psychological particular, consists of a computational system and a lexicon. The computational system determines the structure of linguistic expressions and the subtle interconnections that enter into their form and meaning; the lexicon provides the specific items that appear within them — *tree* in English, *Baum* in German, *ets* in Hebrew, etc. Work of the past several years suggests that the computational system may be fixed and invariant, with the differences among languages reduced to the lexicon, in fact to specific subparts of the lexicon. It may well be that the substantive elements of the lexicon — nouns, verbs, adjectives, and some others — constitute a largely invariant store from which all languages draw, with different decisions as to how they are assigned a physical realization and with the selection from this store determined by the accidental course of experience. Languages differ in certain general features of all lexical items (the best-known case being the option of rightward or leftward assignment of semantic and other properties by lexical items), and in a subpart of the lexicon containing the "grammatical" of "functional" elements such as verbal inflections and case of nouns. Even here, differences may be slight, largely a matter of the specific point at which the computational system links to the perceptual and motor apparatus. Thus the operations of the mind may be uniform across languages though they differ as to which aspects of these operations receive overt expression. English, for example, may well have the nominal inflections of Latin and Greek, but they are not heard, only computed internally, so the evidence suggests.

If this line of thinking is approximately correct, it is close to true that there is only one human language, with slight variations within it depending on accidents of experience. Something like that, in fact, is what any rational Martian scientist would assume, observing the richness and complexity of human language and the poverty of the environment in which it grows and develops in the individual.

The concepts that "grow in the mind" in the virtual absence of determining experience have a rich texture and link together in complex ways. Along with the mechanisms of the computational system, they provide a wide array of semantic connections and an a priori framework for expression of thought in natural language. Included, as a special case, are many kinds of analytic statements, truths of meaning. Some concepts may be, in effect, type variables in the lexicon of natural language, their values more fully determined in other cognitive systems. In these other systems, we may indeed find properties of "meaning holism" and reliance on social practice as commonly alleged, though whether "revision can strike anywhere" is a matter of fact, not to be determined by stipulation. We may proceed, by conscious choice, to abandon natural language and belief systems and to turn to constructions of the rational faculty with quite different properties, stipulating that here nothing is immune to revision; but the fact that we may choose to employ systems in which statements face experience "as a corporate body," as alleged in much current thought about scientific practice, tells us little about the nature of the language faculty or the innate conditions that enter into determining our concepts and systems of belief. Without relevant evidence, a child knows that a box or a ball is an object with strange identity conditions and including its inner space; thus a marble that is in a box is not near the box, because the interior is part of the box in some curious abstract sense, unaffected by whether the box is filled with air or with cheese. We may choose to construct geometrical concepts such as cube and sphere instead of box and ball, so that the marble in the cube may be nearer or farther from it; but such decisions tell us nothing directly about the space of concepts constructed automatically by the mind. The same reasoning holds throughout. Note that for these reasons, we cannot adopt the logic suggested by Hilary Putnam, asking whether some property (holism, or whatever) holds of "the hardest case"; if the "hardest case" is drawn from scientific practice, as in his examples, it entails nothing about ordinary concepts — unless, of course, the rejection of modularity can be sustained.[4]

From this individualistic point of view, to say that a person has a language is to say that some generative procedure is represented in the mind/brain, in fact, a specific modification of the innately determined and largely invariant procedure. A person who has a language, in this sense, knows many things — for example, that certain expressions are pronounced in certain ways, not others, and have certain meanings, not

others. Shall we say, in this case, that the person not only *has* a language
but also *knows* the language? And if so, what is the language that the
person knows? At this point, we enter areas of epistemology that are
confused and poorly understood.

In ordinary discourse, we do say that there is a language, outside of the
mind/brain, which an individual has or knows. Accordingly, we say that
Mary, a five-year-old, has only *partial knowledge* of her language English,
because she hasn't yet learned the words "tendentious" and "sententious";
and Jones, an adult, may have only partial knowledge of his language
English for such reasons and may even be mistaken about his language
because he believes that the word "disinterested" means *uninterested*,
while authorities tell us that it means *unbiased*. We also say that just as
Jones and Smith see the same tree, so they grasp the same meanings,
which are therefore external to their mind/brains. In a more technical
vein, we may say that there is a set of possible languages, determined by
the initial state of the language faculty, and Jones stands in a certain
external relation to one of these languages, the one he has acquired. Jones
and Smith share one of these possible languages, English, which is why
they can understand one another. This "externalized language" that Jones
and Smith share must be an abstract object of some sort, a property of the
community, perhaps. Some have gone on to argue that the concept of
language as a community property, in this sense, is the fundamental one,
and that any other notion, such as the individual sense that I have been
discussing, must be derivative from it (if it is indeed even tenable, as many
deny).

Michael Dummett is one distinguished proponent of this view. He holds
(Dummett 1986) that we must take "the fundamental notion of a language" to be "that of a common language," something like English or
Russian or one of their dialects. Then, he adds, we "have to acknowledge
the partial, and partly erroneous, grasp of the language that every individual speaker has." This common language is a social practice, existing
"independently of any particular speakers," and a word has a meaning in
the common language independently of what some speaker perhaps
"wrongly takes it to mean in the common language." Only on these
assumptions, he argues, can we make sense of the fact that he, Dummett,
does not know Yoruba and knows what he would have to do to acquire
such knowledge, that Turkey suppresses Kurdish, and so on.

A picture of roughly this sort is very widely accepted, and in fact, is
implicit in most of the general discussion about language and thought
among philosophers, linguists, psychologists, and others, and of course in
common sense discourse.

If this picture makes any sense, then it would also make sense to say
that a person who *has* a language in the individualist sense of my remarks
so far — that is, who has a generative procedure represented in the mind/

brain — also *knows* or at least *partially knows* some other language. The difference between *having* and *knowing* is reasonably clear in other cases. Thus Jones may have a moral code or a style of dress; having these, he determines that it is wrong to steal or to wear jeans with a tie. To say further that Jones *knows* a moral code and style of dress is to suppose that there is something outside of Jones's mind/brain, some community property perhaps, that Jones grasps. We can imagine cases in which this move is reasonable; for example, if the moral code or style of dress is presented somewhere in a more or less explicit form, we may say that Jones has a cognitive grasp of that external object. Perhaps some sense can be made of a moral code or style of dress shared implicitly in a community; certainly we assume so, in ordinary discourse. Is it correct to make such moves for the purposes of rational inquiry into the nature of human thought and action, specifically, in the study of human language?

Here skepticism is in order. It is a striking fact that despite the constant reliance on some notion of "community language" or "abstract language," there is virtually no attempt to explain what it might be. In fact, I know of only one attempt to face the problem, by the British philosopher Trevor Pateman (1987). He proposes that we think of the common language of a community as "an (intentional) object of (mutual) belief, appropriately studied hermeneutically within a sociology of language." His characterization of Jones's language in this sense proceeds in terms of Jones's beliefs about his linguistic capacities and behavior and those of others. It is very doubtful that this account — or any like it — captures a real object of the real world, psychological or social. People establish bonds of community in all sorts of intersecting ways, have all sorts of connections with others and beliefs about them, and about themselves. It is far from clear that there is a coherent notion here, even that two individuals will ever share a language, in these terms, given their transient and fluctuating beliefs and associations. One might argue that this is just a normal case of "open texture" and that suitable idealization may allow us to proceed further. That too is doubtful. If we range people by height and weight, we will find some closer to others, but there are no objective categories of "tall," "short," "heavy," and "light," or any reasonable idealizations to be constructed. Communities and the linguistic practices of their members have much the same character, as far as is known, except for the digital character of the array of possibilities, a matter not relevant here. For the present, there is no reason to believe that these are coherent notions, at least for the purposes of theoretical understanding.

Furthermore, even if some notion of shared language can be developed along these lines, it is unclear what is the point of the exercise. For the inquiry into the nature of language, or language acquisition and change, or any of the topics of linguistic inquiry, the notion would appear to have no use — even for sociolinguistics, if we treat it seriously. The notion of "community language" and "community norms" is commonly presupposed as if it is clear enough; it is not, and if it is abandoned, as it probably should be, a great deal of discussion about language, thought, and action will disappear along with it.

Turning to the common sense account of language and its use, borrowed much too uncritically by many theorists, consider the fact that Jones understands Smith when the latter uses the word "tree" to refer to trees. Does it follow that Jones and Smith grasp the same meaning, an object of the common or abstract language? If so, then we should draw the analogous conclusion about pronunciation, given that Jones understands Smith to be saying "tree"; since Jones understands Smith, it must be that there is some object of the common language, the real or common pronunciation of "tree," that Jones and Smith both grasp. No one is inclined to make that move. Rather, we say that Jones and Smith have managed a mutual accommodation that allows Jones, sometimes at least, to select an expression of his own language that for the purposes at hand, matches well enough the one that Smith has produced. There is no need to proceed to the absurd conclusion that there is a common pronunciation that Smith and Jones share or share in part, with a "partially erroneous grasp" in Dummett's sense. The same approach suffices in the case of meaning.

In fact, it is a useful procedure to try to restate standard arguments in the theory of meaning in terms of sound structure and syntax. The reasoning often carries over, no less or more valid, but we are not inclined to make many of the moves that seem plausible — though they are not — in discussing questions of meaning. One can easily be misled by the fact that questions about meaning seem deep and portentous, and are also little understood; the logic is often quite similar to inquiry into domains where these impediments to clear thought do not arise.

Suppose that Smith and Jones have more or less the same shape; we do not conclude that there is a shape that they partially share, and the interactions between Smith and Jones give us no more grounds to suppose that there is a language that they share. There is no general answer to the question of how closely Smith and Jones must match in the steady states of their language faculty for them to communicate, just as there is no answer to the question of how closely they must match in shape to look alike. Dummett doesn't speak Yoruba in the sense that for normal communication, the state of his language faculty is far too remote from the states of the language faculty among a certain group of Africans who match one another closely enough, a highly interest-relative notion; he can change this situation, to a degree, by familiar means. The same is true of the Turks and Kurds. Throughout, there is no sense in trying to construct "common languages," and no meaningful way to do so, apart from particular and fluctuating purposes; there is no notion to be captured by idealization and sharpening our ideas.

As for the class of possible languages, that is a meaningful construct if we take it to be the class of generative procedures, one or more of which is (or is incorporated in) a steady state of the language faculty, assuming now familiar and valid idealizations. In somewhat the same way, we might speak of the set of possible Smiths, each being a possible person developing from Smith's genetic endowment as environmental inputs vary; and more abstractly, the set of possible persons, assuming some idealization to a common human genetic endowment. But we do not say that Smith stands in an external relation to one of the possible Smiths or possible persons; rather, he *is* one of these. Similarly, we need not assume a relation of Smith's language faculty in its steady state to one of the possible languages (taken as generative procedures); rather, Smith's language faculty *is* (or incorporates) one of these possible languages. Talk of "norms" and "conventions" faces similar problems, and it is unclear what if anything will remain of the analyses that rely on these notions, if the problems are faced instead of mistakenly assumed to be unimportant.

Similar questions arise about the notion of "partial knowledge" and "misuse." Take Mary, the five-year-old child. Of which language does she have only partial knowledge? We may provide various interest-relative answers, but the only clear and theoretically useful one seems to be: Mary has partial knowledge of any consistent extension of her current generative procedure (where "consistency" is defined relative to the absolute notion of initial state). Similarly, we might say that in the initial state, Mary had "partial knowledge" of all possible languages. As for Jones, who misuses "disinterested" (perhaps, in the manner of everyone to whom he relates in some "community"), the observation makes as much sense as the statement that Jones is mispronouncing the word when he speaks in his own variety of what we loosely call "English," or is misusing French when he speaks German. If there is anything sensible to say about these topics apart from consideration of colors on maps, authority structures, and the like, it has yet to be discovered. There are crucial and justified idealizations throughout the study of language or any other real-world phenomenon, but not, in seems, in these domains. There is no argument to be found here against strictly internalist approaches that take a language to be a psychological particular.

In a thoughtful review of recent debates over externalist and internalist approaches to the theory of meaning and thought, Akeel Bilgrami (1989) argues plausibly that behind the specific arguments about twin-earth problems and the like, there lies a deeper concern: assuming "that thought and meaning must be public phenomena . . . , then the following is a good question: How shall we characterize thought and meaning such that its public availability is ensured?" Internalist approaches, regarding thought and meaning as "private," leave this a mystery; but "If another's meanings and propositional attitudes are determined by items in a world external to her, then it is neither surprising nor avoidable that they are available to one who lives in the shared environment." Similar considerations may underlie the commitment to a public "common language," and to the social character of language generally. Thus in arguing in favor of what he has called "the division of linguistic labor," Hilary Putnam rejects certain internalist proposals as to the meaning or "personal concept" of, say, the term "elm," on grounds that they do not permit "elm" to be translated into German "Ulme," and there would be "immense difficulty" in translating a host of other common nouns. The problem, he argues, is to account for the public nature of meaning and thought, and it dissolves if we assume that "reference is a social phenomenon," not an individual one.[5]

There are at least three problems with such arguments. First, no useful sense has been given to the notion of "social phenomenon," and it is doubtful that any can be; for example, what is the relevant "community" and "community language" if the expert to whom I defer to determine the reference of "elm" happens to be an Italian gardener with whom I share only certain Latin names? Second, if there are general problems about the public nature of meaning and thought (which is far from clear), it seems that these would become more difficult insofar as reference is a social phenomenon, understood differently by different speakers with their "partial grasps" of the alleged truth. Furthermore, it has to be established that translation is anything more than a search for the best match for some purpose, and that thought and meaning really are public phenomena, except in roughly the (highly misleading) sense in which pronunciation and shape are public phenomena. No such dubious assumptions are needed to account for communication or other known phenomena.

Bilgrami takes note of the possibility that the public nature of thought and meaning might be denied, but holds that externalism is still defensible on other grounds. He argues that internalists like Descartes have no way to justify their "right . . . to concepts of objective and external things" such as tables and chairs, while the externalist faces no such problem, because our experience and thought are often "the experience of objective and external things," and indexical element in thought and meaning. But the consistent internalist has no need to justify the concepts he has, any more than he justifies his pronunciation or, for that matter, his circulatory system. The concepts simply grow in the mind/brain in the way they do, and are then available for use. As for the fact that particular concepts develop in the mind on the basis of experience, not others, that is a reflection of the initial state of the mind/brain, and the externalist and internalist appear to face the same problems in accounting for it.

It may be that a kind of public character to thought and meaning is guaranteed by the near uniformity of initial endowment, which permits only the attainment of steady states that vary little in relevant respects. Beyond this, their character varies as experience varies, with no clear way to establish further categories, even ideally. And to the extent that uniformity across the species exists, it is surely no necessity, but only a matter of contingent fact, not relevant to the externalist thesis.

It seems, then, that the only sense of language we need consider is *generative procedure*; to have a language is to have such a procedure, represented in the mind/brain. Having a language, Jones has a way of speaking and understanding. We may say, in these terms, that he knows that so-and-so means such-and-such and is pronounced in such and such a way, relative to his generative procedure; there being no external reality against which to compare his judgments, we have a rather special case of propositional knowledge (also, knowledge how, why, of, etc.). Though far less is understood about the matter, it seems reasonable to adopt a similar approach in the case of the cognitive structures that enter into moral judgment, functioning in a social community, and much else, in fact, in any domain where the problem of poverty of stimulus arises.

As for the rational faculty, while recognizing the qualitative differences already noted, nevertheless there is much that is the same. Take a "problem situation" to be determined by some state of understanding, some array of phenomena subjected to inquiry, and some questions formulated about them. Typically, the abductive faculties provide no intelligible theories, but sometimes they do yield one or a small set that our rational faculty considers close enough to the phenomena to merit further inquiry. Although these theories, in interesting cases, are radically underdetermined by the problem situation in the usual sense, nevertheless there is often substantial convergence among researchers. Whether we are speaking of Kuhn's normal science or the occasional revolutions, the course pursued suggests that the rational faculty, like the language faculty, reaches a new state in a manner determined by its nature, however this may be refined by the procedures of testing and confirmation, which also reflect our intuitive concepts of rationality. The mind, then, "grows theories," though in quite different ways than in other cases we partially understand, a fact that should not be surprising.

So far, we have kept to the framework of what is sometimes called "naturalized epistemology," though without the dogmatic and untenable framework of conditioning and the like — and in truth, without saying a great deal, because so little is known about the rational faculty and its character. Epistemology naturalized in this sense is in the spirit of much of classical epistemology. The cognitive structures that are developed in this way may be given a more or less explicit characterization, in which case we call them "science." In rare cases, it is even worthwhile to proceed to the level of true formalization, as in parts of modern mathematics and physics where understanding can be advanced by rigorous proof of theorems and precise derivation of consequences that are not at once apparent. Given such cognitive structures, we try to determine whether they capture important features of the real world that lies beyond our direct apprehension, in accord with the stance of constructive skepticism; if they do, we regard the theories expressed as at least partially true, and we say that we have achieved a degree of knowledge of the world, or knowledge that such-and-such, and so on. Growth of knowledge may involve "advancement of understanding," not only "fixation of belief," as Goodman observes in a different connection.[6] There is a good deal more to say if we want to capture the term "knowledge" as commonly used, but it may be that the essential tasks of classical or naturalized epistemology largely fall within this framework. If so, then coming to "have a language" may be an informative example of what is involved in attainment of understanding and knowledge in other domains, despite the absence, in this case, of a presumed external object against which the constructions of the mind can be compared for accuracy.

Questions of this nature came to the fore in the context of what some call "the cognitive revolution" of the 1950s, which shifted perspective from behavior and its products to the inner mechanisms of the mind/brain that enter into action and interpretation, and thus directed the study of these topics towards the core natural sciences. To an extent not then appreciated, these developments recapitulated steps taken, often with much insight, in what we might call "the first cognitive revolution" of the 17th century, which had particularly important achievements in the study of vision and language, the two areas that have progressed the most in the second cognitive revolution. The basic questions posed are those of cognitive structure and cognitive growth, their determinants in our biologically-determined nature, the question of how the mind "makes worlds," as it develops on a course that appears to be internally directed in significant ways.

In employing the rational faculty, in part a conscious and controlled procedure, we are attempting to enhance understanding. But the process of making worlds may have quite different motives and social functions. Its explicit goal or tacit social function, institutionally maintained, may be to prevent understanding. Here we enter the domain of Orwell's problem.

Like the questions of cognitive growth and structure, the problem of indoctrination and social control also loomed large in the 17th century. During the English revolution, libertarian groups "represented the first great outburst of democratic thought in history," with the Levellers leading the way, one historian comments.[7] This outburst of democratic thought at once raised the problem of how to contain the threat, an elite reaction that has had a good deal of resonance since. The libertarian ideas of the radical democrats were considered outrageous by respectable people. They favored universal education, guaranteed health care, and democratization of the law, which one described as a fox, with poor men the geese: "he pulls off their feathers and feeds upon them." They developed a kind of "liberation theology" which, one critic ominously observed, aimed "to preach anti-monarchical seditious doctrine to the people . . . , to raise the rascal multitude and schismatic rabble against all men of best quality in the kingdom, to draw them into associations and combinations with one another in every county and with the Army, against all lords, gentry, ministers, lawyers, rich and peaceable men" (historian Clement Walker). The rabble did not want to be ruled by King or Parliament, but "by countrymen like ourselves, that know our wants." Their pamphlets explained that "It will never be a good world while knights and gentlemen make us laws, that are chosen for fear and do but oppress us, and do not know the people's sores."

These ideas naturally appalled the men of best quality. They were willing to grant the people rights, but within reason, and on the principle that "when we mention the people, we do not mean the confused promiscuous body of the people." Particularly frightening were the itinerant preachers and mechanics preaching freedom and democracy, the agitators stirring up the rascal multitude, and the printers putting out pamphlets questioning authority and its mysteries. "There can be no form of government without its proper mysteries," Clement Walker warned, mysteries that must be "concealed" from the common folk. In words echoed by Dostoevsky's Grand Inquisitor, he went on to observe that "Ignorance, and admiration arising from ignorance, are the parents of civil devotion and obedience." The radical democrats had "cast all the mysteries and secrets of government . . . before the vulgar (like pearls before swine)," he continued, "and have taught both the soldiery and people to look so far into them as to ravel back all governments to the first principles of nature. . . . They have made the people thereby so curious and so arrogant that they will never find humility enough to submit to a civil rule." It is dangerous, another commentator ominously observed, to "have a people know their own strength." After the democrats had been defeated, John Locke wrote: "day-labourers and tradesmen, the spinsters and dairymaids" must be told what to believe; "The greatest part cannot know and therefore they must believe."[8]

Like John Milton and other civil libertarians of the period, Locke held a sharply limited conception of freedom of expression, barring those who "speak anything in their religious assembly irreverently or seditiously of the government or governors, or of state matters." The common people should be denied the right to discuss public affairs; Locke's Fundamental Constitution of Carolina provided that "all manner of comments and expositions on any part of these constitutions, or on any part of the common or statute laws of Carolines, are absolutely prohibited." In drafting reasons for Parliament to terminate censorship in 1694, Locke offered no defense of freedom of expression or thought, but only considerations of expediency and harm to commercial interests.[9] The threat of democracy having been overcome with the defeat of the libertarian rabble, censorship was permitted to lapse in England, because the "opinion-formers . . . censored themselves. Nothing got into print which frightened the men of property," Christopher Hill observes.

The concerns aroused by the 17th century radical democrats were not new. As far back as Herodotus we can read how people who had struggled to gain their freedom "became once more subject to autocratic government" through the acts of able and ambitious leaders who "introduced for the first time the ceremonial of royalty," distancing the leader from the public while creating a legend that "he was a being of a different order from mere men" who must be shrouded in mystery, and leaving the secrets of government, which are not the affair of the vulgar, to those entitled to manage them.

Such problems regularly arise in periods of turmoil and social revolution, when it is imperative to ensure that the people are driven back to a proper humility, and that the better kind of people rule, whether they are called Bolsheviks, democrats, or whatever. During the American revolution, rebellious and independent farmers had to be taught, by force, that the ideals expressed in the pamphlets of 1776 were not to be taken seriously. The common people were not to be represented by countrymen like themselves, that know the peoples' sores, but by gentry, merchants, lawyers, and others who hold or serve private power. The reigning doctrine, expressed by the Founding Fathers, is that "the people who own the country ought to govern it," in John Jay's words. The rise of corporations in the 19th century, and the legal structures devised to grant them dominance over private and public life, established the victory of the Federalist opponents of popular democracy in a new and powerful form.

Quite regularly, revolutionary struggles pit aspirants to power against one another though united in opposition to radical democratic tendencies among the common people. Lenin and Trotsky, shortly after seizing state power in 1917, moved to dismantle organs of popular control, including factory councils and Soviets, thus to deter and overcome any socialist tendencies. An orthodox Marxist, Lenin did not regard socialism as a viable option in this backward and underdeveloped country. In what has always seemed to me his greatest book, Orwell described similar developments in Spain, where the fascists, Communists, and liberal democracies were united in opposition to the popular libertarian revolution that swept much of the country, turning to the struggle over the spoils only when it was suppressed. There are many other examples, often crucially influenced by great power violence.

Just before his assassination, Archbishop Romero of San Salvador pleaded in vain with President Carter to withhold aid from the murderous military junta who are seeking to destroy the "peoples' organizations fighting to defend their fundamental human rights." Such popular organizations are perceived as an intolerable threat by the men of best quality. That is why there is near unanimity that they must be crushed by force, as was done in this case by U.S.-organized state terror in the style of Pol Pot, with the tacit support of the western democracies. To mention only Holland, after the security forces had murdered the political opposition, destroyed the independent media by violence, occupied and devastated the university with many killed, decapitated the unions and popular organizations, and slaughtered tens of thousands of people leaving piles of bones and dismembered and mutilated corpses to intimidate the survivors, Washington organized "elections" to legitimate the terror. The Dutch government sent observers, who wrote in their official report that although "the parties of the left were excluded to a certain extent from the election process" (meaning, they were slaughtered and driven into exile), nevertheless "there was a sufficient range of choices for the voters."[10] Comparable apologetics for the crimes of some official enemy are unthinkable, and if they were to be voiced, the horror and revulsion would be uncontained. I would be pleasantly surprised to learn that was true in the present case.

The reason for the fear of popular organizations is that they might lay the basis for meaningful democracy and social reform, thus threatening the prerogatives of the privileged. Worse still, "the rot may spread," in the terminology of U.S. government planners since World War II; there may be a demonstration effect of successful independent development in a form that attends to the peoples' sores.

Similar fears were expressed by European guardians of order with regard to the American revolution, which might "lend new strength to the apostles of sedition," Metternich warned; it might spread "the contagion and the invasion of vicious principles" such as "the pernicious doctrines of republicanism and popular self-rule," one of the Czar's diplomats explained. A century later, the cast of characters was reversed. Woodrow Wilson's Secretary of State Robert Lansing warned that the Bolshevik disease, were it to spread, would leave the "ignorant and incapable mass of humanity dominant in the earth"; the Bolsheviks, he continued, were appealing "to the ignorant and mentally deficient, who by their numbers are urged to become masters, . . . a very real danger in view of the process of social unrest throughout the world"; it is, as always, democracy that is the awesome threat. When soldiers and workers councils made a brief appearance in Germany, Woodrow Wilson feared that they would inspire dangerous thoughts among "the American negro [soldiers] returning from abroad." Already, negro laundresses were demanding more than the going wage, saying that "money is as much mine as it is yours," Wilson had heard. Businessmen might have to adjust to having workers on their boards of directors, he feared, among other disasters if the Bolshevik virus were not exterminated.

With these dire consequences in mind, the Western invasion of the Soviet Union was justified on defensive grounds, in defense against "the Revolution's challenge . . . to the very survival of the capitalist order," its "self-proclaimed intention to seek the overthrow of capitalist governments throughout the world." This is the retrospective judgment after 70 years by the leading diplomatic historian John Lewis Gaddis, who justifies the Western intervention to overthrow the Soviet government in these "defensive" terms in a highly-regarded recent book.[11]

Given these perceptions, it was only natural that the defense of the United States should extend from invasion of the Soviet Union to Wilson's Red Scare at home. As Lansing explained, force must be used to prevent "the leaders of Bolshevism and anarchy" from proceeding to "organize or preach against government in the United States"; the government must not permit "these fanatics to enjoy the liberty which they now seek to destroy," specifically, the liberty of the rich and powerful to rule and exploit. The repression launched by the Wilson administration successfully undermined democratic politics, unions, freedom of the press, and independent thought, in the interests of corporate power and the state authorities who represented its interests, all with the general approval of the media and elites generally, all in self-defense against the ignorant rabble.

The themes persist to the present. In the 1970s, the Trilateral Commission, bringing together liberal elites from Europe, Japan, and the United States, warned of an impending "crisis of democracy" because the "excess of democracy" was posing a threat to the unhampered rule of privileged elites — what is called "democracy" in political theology. The problem was that the rabble were attempting to enter the political arena to press their demands after the ferment of the sixties, with its cultural awakening and organizing efforts among young people, ethnic minorities, women, social activists, and others, encouraged by the struggles of benighted masses elsewhere for freedom and independence. More "moderation in democracy" would be required, the Commission concluded, a return to the good old days when "Truman had been able to govern the country with the cooperation of a relatively small number of Wall Street lawyers and bankers," as the American rapporteur commented with a trace of nostalgia.[12]

At another point on the political spectrum, the conservative contempt for democracy is succinctly articulated by Sir Lewis Namier, who writes that "there is no free will in the thinking and actions of the masses, any more than in the revolutions of planets, in the migrations of birds, and in the plunging of hordes of lemmings into the sea." Only disaster would ensue if the masses were permitted to enter the arena of decision-making in a meaningful way. The leading neo-conservative intellectual Irving

Kristol caustically observes (Kristol 1973) that "insignificant nations, like insignificant people, can quickly experience delusions of significance." These delusions must be driven from their tiny minds by force, he continues: "In truth, the days of 'gunboat diplomacy' are never over ... Gunboats are as necessary for international order as police cars are for domestic order." Note that Kristol's basic conceptions changed little as he shifted from Trotskyite militant to servant of corporate power, a conventional transition as assessments of where true power lies are modified by experience.

Continuing with what is called "conservatism" in one of the many recent Orwellian perversions of political discourse, we may move on to the Reagan administration, which established a state propaganda agency that was by far the most extensive in American history, dedicated to mobilizing support for the U.S. terror states in Central America and to "demonizing the Sandinistas," in the words of a high administration official. When the program was exposed, one high official described it as the kind of operation carried out in "enemy territory," an apt phrase, expressing standard elite attitudes towards the public: an enemy, who must be subdued.

As usual, fear of the domestic public went hand-in-hand with concerns over the spreading of the rot and the cancers abroad. One of the most spectacular achievements of this state propaganda operation, strictly illegal as Congress irrelevantly determined, was to concoct the charge that Nicaragua was planning to conquer the hemisphere. The proof was that the Sandinistas had declared a "Revolution without Borders." This charge — which aroused no ridicule among the disciplined educated classes — was based on a speech by Sandinista leader Tomás Borge, in which he explained that Nicaragua cannot "export our revolution" but can only "export our example" while "the people themselves of these countries ... must make their revolutions"; in this sense, he said, the Nicaraguan revolution "transcends national boundaries." The fraud was at once exposed, but was eagerly accepted by Congress, the media, and political commentators, and has been exploited since as the justification for maintaining what the World Court declared to be the "unlawful use of force" and illegal violation of treaties in Washington's war against Nicaragua. Observe that lying behind the fraud is a valid insight, which explains its virtually unanimous appeal among the educated classes. Early Sandinista successes in instituting social reforms and production for domestic needs, highly praised by the international lending agencies and relief and development organizations, set the alarm bells ringing in Washington and New York, arousing the same fears that inspired Metternich and the Czar, the people of best quality since the 17th century, all those who expect to dominate by right: the rot might spread, and the foundations of privilege might crumble.

We may note that the furor over the Sandinista threat to carry out a "revolution without borders" proceeds side-by-side with applause for the U.S. effort to impose its preferred model by violence. The same journals that denounce the Sandinistas, cheerfully parroting Washington propaganda about "revolution without borders" that they know to be fraudulent, publish without a second thought effusive praise for a book subtitled "exporting the American revolution" and advocate forceful intervention to achieve this result.[13] The *actual* use of force to export the U.S. model, in short, is praiseworthy. But the Nicaraguan attempt to construct a society that others will attempt to emulate arouses passionate anger, even the declaration of a National Emergency in the United States in 1985, since renewed annually, most recently in April of this year (1989), because "the policies and actions of the government of Nicaragua continue to pose and unusual and extraordinary threat to the national security and foreign policy of the United States" (George Bush).[14] This is not, of course, empty rhetoric. These hysterical pronouncements provide the legal basis for the embargo that serves to strangle Nicaragua and to maximize suffering, starvation, and disease along the lines preferred by the liberal doves who explain that we must "contain Nicaragua" and restore it to the "Central American mode" exemplified by the U.S. terror states.

Concern that "the rot may spread" is a dominant factor in state planning, as extensively documented in the secret and public records, not to speak of the ample testimony of history. But these matters are largely suppressed by the loyal intellectual classes in their own service to order and good form. It is, after all, important to ensure that the "ignorant and mentally deficient" common people not comprehend the workings of power and privilege — the fundamental principle behind Orwell's problem.

The rabble, however, continue to fight for their rights, and over time, the libertarian ideals that often receive some form of spontaneous expression during periods of turmoil and conflict have been partially realized or have even become common coin. Many of the outrageous ideas of the 17th century radical democrats, for example, seem tame enough today, though other insights remain beyond our current moral and intellectual reach.

The struggle for freedom of speech is an interesting case. "The central problem of free speech," legal historian Harry Kalven comments, "is to determine when, if ever, the *content* of communications may be interdicted by law." One critical element is seditious libel, the idea that the state can be criminally assaulted by speech, "the hallmark of closed societies throughout the world," Kalven observes. A society that tolerates seditious libel is not free, whatever its other characteristics, he continues. In late 17th century England, men were castrated, disemboweled, quartered and beheaded for the crime. Through the 18th century, Leonard Levy observes, Englishmen "resolutely continued to ride the crest of an early consensus that the maintenance of established authority demanded the silencing of subversive discussion" and that "any threat, whether real or imagined, to the good reputation of the government" must be barred by force. "Private men are not judges of their superiors . . . [for] This wou'd confound all government," one editor wrote. Men of "post and quality" must be protected from charges, the true being more criminal than the false, or "they cou'd not govern, nor wou'd they be obey'd as they ought to be if they were render'd contemptible to their subjects, which is unavoidable if they are suffered to be traduced by every private person, and expos'd all over the nation."

The same principles were upheld in the American colonies, despite vibrant and lively debate. The intolerance of dissent during the revolutionary period is notorious. The leading American libertarian, Thomas Jefferson, agreed that punishment was proper for "a traitor in thought, but not in deed," and authorized internment of political suspects. It was not until the Jeffersonians were themselves subjected to harsh and repressive measures in the late 1790s that they developed a body of more libertarian thought for self-protection — reversing course, however, when they gained power themselves.[15]

Until World War I, there was only a slender basis for freedom of speech in the United States, and it was not until 1964 that the law of seditious libel was struck down by the Supreme Court. In 1969, the Court finally protected speech apart from "incitement to imminent lawless action," thus formulating a libertarian standard which, I believe, is unique in the world. In Canada, for example, people are still imprisoned for promulgating "false news," established as a crime in 1275 to protect the King. In Europe, to my knowledge, the situation is in some ways still more primitive, and a close look at the practices of the intellectual community there is less than edifying, to put it rather charitably.

We should also bear in mind that the right to freedom of speech in the United States was not established by the First Amendment to the Constitution, but only through committed efforts over a long term by the labor movement and the civil rights and anti-war movements of the 1960s, and other popular forces. As James Madison observed, a "parchment barrier" will never suffice to prevent tyranny. Rights are not established by words, but won and sustained by struggle.[16]

It is also worth recalling that the victories for freedom of speech are often won in defense of the most depraved and horrendous views. The 1969 Supreme Court decision that established a unique libertarian standard was in a case defending the Ku Klux Klan from prosecution after a meeting with hooded figures, guns, and a burning cross, calling for "burying the nigger" and "sending the Jews back to Israel."[17] Among the mullahs in Qom — and, unfortunately, much closer to where we meet — it is necessary to stress these simple matters.

The fears expressed by the men of best quality in the 17th century have become a major theme of intellectual discourse, corporate practice, and the academic social sciences in the 20th century. The influential moralist and foreign affairs adviser Reinhold Niebuhr, revered by George Kennan, the Kennedy intellectuals, and many others, wrote that "rationality belongs to the cool observers" while the common person follows not reason but faith. The cool observers, he explained, must recognize "the stupidity of the average man," and must provide the "necessary illusions" and the "emotionally potent oversimplifications" that will keep the naive simpletons on course. When he wrote these words in the early 1930s, Niebuhr was in his leftist phase. In this case too, the basic conceptions underwent little change as he passed through the conventional "God that failed" transition, taking on the role of "official establishment theologian" in the admiring words of political commentator Richard Rovere, offering counsel to those who face the responsibilities of power.[18]

In accordance with the prevailing conceptions, there is no infringement on democracy if a few corporations control the information system: in fact, that is the essence of democracy. In the *Annals of the American Academy of Political and Social Science*, the leading figure of the public relations industry, Edward Bernays, explained that "the very essence of the democratic process" is "the freedom to persuade and suggest," what he calls "the engineering of consent." "A leader," he continues, "frequently cannot wait for the people to arrive at even general understanding . . . Democratic leaders must play their part in . . . engineering . . . consent to socially constructive goals and values," applying "scientific principles and tried practices to the task of getting people to support ideas and programs"; and although it remains unsaid, it is evident enough that those who control resources will be in a position to judge what is "socially constructive," to engineer consent through the media, and to implement policy through the mechanisms of the state. If the freedom to persuade happens to be concentrated in a few hands, we must recognize that such is the nature of a free society. The public relations industry expends vast resources "educating the American people about the economic facts of life" to ensure a favorable climate for business. Its task is to control "the public mind," which is "the only serious danger confronting the company," an AT&T executive observed eighty years ago. And today, the *Wall Street Journal* describes with enthusiasm the "concerted efforts" of corporate America "to change the attitudes and values of workers" on a vast scale with "New Age workshops" and other contemporary devices of indoctrination and stupefaction designed to convert "worker apathy into corporate allegiance."[19] The agents of Reverend Moon and Christian evangelicals employ similar devices to bar the threat of peasant organizing and to undermine a church that serves the poor in Latin America. They are amply funded for these activities by the intelligence agencies of the U.S.

and its clients and the closely-linked international organizations of the ultra-right.

Bernays expressed the basic point with lucidity in a public relations manual of 1928: "The conscious and intelligent manipulation of the organized habits and opinions of the masses is an important element in democratic society. Those who manipulate these unseen mechanisms of society constitute an invisible government which is the true ruling power of our country . . . It is the intelligent minorities which need to make use of propaganda continuously and systematically."[20] The intelligent minorities have long served this function, often with awareness and self-justification: it is all done in the interests of the ignorant rabble, who will only cause trouble if they are given a free hand.

Such ideas are common across the political spectrum. The dean of U.S. journalists, Walter Lippmann, described a "revolution" in "the practice of democracy" as "the manufacture of consent" has become "a self-conscious art and a regular organ of popular government." This is a natural development when "the common interests very largely elude public opinion entirely, and can be managed only by a specialized class whose personal interests reach beyond the locality." He was writing shortly after World War I, when the liberal intellectual community was much impressed with its success in serving as "the faithful and helpful interpreters of what seems to be one of the greatest enterprises ever undertaken by an American president" (*New Republic*). The enterprise was Woodrow Wilson's interpretation of his electoral mandate for "peace without victory" as the occasion for pursuing victory without peace, with the assistance of the liberal intellectuals, who later praised themselves for having "impose[d] their will upon a reluctant or indifferent majority," with the aid of propaganda fabrications about Hun atrocities and other such devices. They were serving, often unwittingly, as instruments of the British Ministry of Information, which secretly defined its task as "to direct the thought of most of the world."[21]

Fifteen years later, Harold Lasswell explained in the *Encyclopaedia of the Social Sciences* that we should not succumb to "democratic dogmatisms about men being the best judges of their own interests." They are not; the best judges are the elites, who must, therefore, be ensured the means to impose their will, for the common good. When social arrangements deny them the requisite force to compel obedience, it is necessary to turn to "a whole new technique of control, largely through propaganda" because of the "ignorance and superstition [of] . . . the masses."

These doctrines are entirely natural in any society in which power is narrowly concentrated but formal mechanisms exist by which ordinary people may, in theory, play some role in shaping their own affairs — a threat that plainly must be barred.

The techniques of manufacture of consent are most finely honed in the

United States, a more advanced business-run society than its allies and
one that is in many ways more free than elsewhere, so that the ignorant
and stupid masses are potentially more dangerous. But the same concerns
are common in Europe, as already illustrated. In August 1943, Jan
Christiaan Smuts warned his friend Winston Churchill that "with politics
let loose among those peoples, we may have a wave of disorder and
wholesale Communism set going all over those parts of Europe." Churchill's conception was that "the government of the world" should be in the
hands of "rich men dwelling at peace within their habitations," who had no
"reason to seek for anything more" and thus would keep the peace,
excluding those who were "hungry" and "ambitious." The same precepts
apply at home. Smuts was referring specifically to southern Europe,
though the concerns were far broader. With conservative elites discredited
by their association with fascism and radical democratic ideas in the air, it
was necessary to pursue a worldwide program to crush the anti-fascist
resistance and its popular base and to restore the traditional order, to
ensure that politics would not be let loose among those peoples; this
campaign, conducted from Korea to western Europe, would be the topic
of the first chapter of a serious work on post-World War II history.[22]

The same problems arise today, heightened, in Europe, by the fact that
unlike the United States, its variety of state capitalism has not yet largely
eliminated labor unions, excluded political parties apart from factions of
the business party, and barred other impediments to rule by people of the
best quality. These persistent concerns help explain the ambivalence of
European elites towards détente, with the concomitant decline of the
technique of social control through fear of the great enemy.

The basic problem, as expressed by Harold Lasswell and many others,
is that as the state loses the capacity to control the population by force
and violence, privileged sectors must find other methods to ensure that the
public is marginalized and removed from the public arena. And the
insignificant nations must be subjected to the same practices as the
insignificant people. The dilemma was explained by Robert Pastor, Latin
American specialist of the Carter Administration, at the extreme liberal
and dovish end of the political spectrum. Defending U.S. policy over many
years, he writes that "the United States did not want to control Nicaragua
or other nations in the region, but it also did not want to allow developments to get out of control. It wanted Nicaraguans to act independently,
except when doing so would affect U.S. interests adversely."[23] In short,
Nicaragua and other countries should be free — to do what we want them
to do — and should choose their course independently, as long as their
choice conforms to our interests. If they use the freedom we accord them
unwisely, then naturally we are entitled to respond in self-defense. The
ideas expressed are a close counterpart to the prevailing liberal conception
of democracy at home as a form of population control. At the other

extreme of the spectrum, we find the "conservatives" with their preference for quick resort to Kristol's methods: gunboats and police cars.

The techniques of manufacture of consent have been amply discussed and exemplified elsewhere, primarily with regard to the media. There are by now thousands of pages of documentation confirming the conclusion that the media serve the social function of indoctrination in the interests of state and private power and privilege. This thesis has been subjected to a wide range of critical tests, and has withstood them very well. To my knowledge, it faces no serious challenge, and is well-confirmed by the standards of the social sciences. Inquiry into the elite intellectual culture and mainstream scholarship leads to rather similar conclusions, with some qualifications.

The systematic study of these matters is largely restricted to the United States. This imbalance may lead to the belief that the European media and intellectual culture are more free of the influences of centralized power and privilege, more independent and objective. That is one possibility; another is that the absence of study reflects a more uncritical attitude towards power and privilege on the part of European intellectuals. The few cases that have been studied suggest that both of these conclusions may be valid; note that there is no inconsistency. The political spectrum is wider in Europe, and this is reflected in the media, scholarship, and other domains. On the other hand, the intellectual environment is often less critical and skeptical, despite a very different self-image. One very carefully executed study of coverage of Salvadoran and Nicaraguan elections by a wide range of European media reveals quite remarkable conformity to the framework dictated by U.S. government propaganda, not quite as extreme as shown in comparable studies of the U.S. media, but highly revealing nonetheless.[24] I have documented cases where French writers who are familiar with the United States have produced arrant nonsense in the service of dominant ideology in articles and books in France (and elsewhere in Europe), assuming correctly that no one would care, though they were careful to introduce crucial corrections in the United States to protect themselves from quick exposure. Concealing of state involvement in major atrocities has also been a simple matter in the European intellectual climate, along with suppression of freedom of press and of speech, which passes unnoticed, or is even praised. There is a great deal more to say about these matters, but I will not proceed with them except to warn against conclusions drawn from failure to investigate.[25]

A properly functioning system of indoctrination has a variety of tasks, some rather delicate. The stupid and ignorant masses must be kept that way, diverted with emotionally potent oversimplifications, marginalized, and isolated. Ideally, each person should be alone in front of the TV screen watching sports, soap operas, or comedies, without the kinds of organizational structure that permit individuals lacking resources to discover what they think and believe in interaction with others, to formulate their own concerns and programs, and to act to realize them. They can then be permitted, even encouraged, to ratify the decisions made by their betters in periodic elections. For those expected to take part in serious decision-making and control, the problem of indoctrination is a bit different. The business, state, and cultural managers, and articulate sectors generally, must internalize the values of the system and share the necessary illusions that permit it to function in the interests of concentrated power and privilege. But they must also have a certain grasp of the realities of the world, or they will be unable to perform their tasks effectively. It is not easy to steer a way through these dilemmas, and it is quite intriguing to see in detail how it is done, but that is beyond the scope of the discussion here.

My personal sense of the matter is that Plato's problem is to be approached in the manner of the sciences, and in some areas, is susceptible to significant progress and achievements. As for Orwell's problem, perhaps it too has unexplored depths, as the preliminary statement for this meeting suggests. Even a cursory look shows how readily our linguistic-conceptual devices lend themselves to presenting the facts of the world in a manner conforming to narrowly-conceived self-interest; there are paired terms that typically refer to actions that are indistinguishable in themselves, but are either *yours* (terrorism and aggression) or *mine* (retaliation and defense), a rhetorical system designed so that whatever happens can be described with positive or negative values, with self-righteous exculpation or vast indignation, as interest demands. There is doubtless more to learn, perhaps along lines sketched in the preliminary statement.

Whether this is true or not, we can be confident that Orwell's problem will be with us until effective power to make the crucial social decisions is diffused, as radical democrats and libertarians have long advocated. There are tendencies in that direction, others that run counter to it. We can only speculate as to whether the libertarian ideals that were crafted through the Enlightenment and since may yet be consummated, beyond the very partial achievements of the past centuries. Until they are, Orwell's problem will surely be a dominant feature of the cultural landscape.

NOTES

[1] Davidson (1986). See also Putnam (1988) for arguments concerning meaning in natural language based on rationality in science; Putnam does offer one argument to support the transition, but it is not strictly relevant, turning on properties of informal reasoning (p. 9). Note that the conclusions about meaning holism, social practice, and the like might be correct for natural language, at least in part, even if the arguments offered do not hold. See below, and for more on the matter, Chomsky (forthcoming).
[2] Popkin (1979, pp. 48, 140f.); Goodman (1984, p. 22).
[3] See the papers in Pinker and Mehler (1988), for a close study of these topics.

[4] See references of note 1.

[5] Putnam, op. cit., p. 26f. As noted earlier, the conclusion may be accurate even if the argument fails. In fact, I think there are reasons to accept this conclusion for subparts of the vocabulary such as natural kind terms, for other reasons, and to redraw the boundaries of what counts as "natural language" vs. systems of thought and belief accordingly; but that is another matter, and it does not impugn the well-established conclusion that natural language induces intrinsic semantic connections (including, in particular, analyticity) and that the thesis of meaning holism is not generally true.

[6] Op. cit.

[7] Margaret Judson, cited by Levy (1985, p. 91).

[8] Hill (1975). With regard to Locke, Hill adds, "at least Locke did not intend that priests should do the telling; that was for God himself."

[9] Levy, op. cit.

[10] Dr Thom Kerstiens and Dr Piet Nelissen (1984), a report on elections in Nicaragua, including comparisons to El Salvador.

[11] For references here and below, where not otherwise cited, see Chomsky (1985, 1986b). For Lansing and Wilson, see Gardner (1987, pp. 157, 161, 261, 242).

[12] For more on these matters, see Chomsky (1977).

[13] Fossedal (1989), advertised in the *New Republic* and elsewhere with endorsements by distinguished figures, including Richard Nixon, who supports the book's message that the U.S. must pursue "an activist, and even interventionist, foreign policy."

[14] AP, April 21, 1989, unpublished, to my knowledge. The media have often not published these statements, perhaps in embarrassment, perhaps to ward off a negative public reaction.

[15] Levy, op. cit., pp. 178—179, 297, 337ff.

[16] Levy, op. cit., pp. xvii, 6, 9, 102; Kalven (1988, pp. 63, 227f.).

[17] Ibid., p. 121f.

[18] On the nature of his counsel and the thinking that has led to extraordinary acclaim, see Chomsky (1987).

[19] Cited by Schiller (1989).

[20] Cited by Preston and Ray (1988).

[21] Cited from secret documents by Marlin (1989). See reference of note 12. For further discussion of Lippmann's doctrines, in a more general context, see now Chomsky (1991, chapter 12, based in part on this lecture).

[22] For some details, see Chomsky (1989a), and sources cited; and the expanded version in Chomsky (1991), chapter 11.

[23] Pastor (1987, p. 32), his emphasis.

[24] Rietman (1988). For more on media coverage of the elections, see Herman and Chomsky (1988, chapter 3), and Chomsky (1988b, pp. 141—142).

[25] On these matters, see Chomsky and Herman (1979), Chomsky (1989b).

REFERENCES

Bilgrami, Akeel: 1989, 'Realism Without Internalism', *Journal of Philosophy* **LXXXVI** (2).

Chomsky, Noam: 1977, 'Intellectuals and the State', Huizinga lecture, Leiden. Het Wereld-venster, Baarn.

Chomsky, Noam: 1982, *Towards a New Cold War*, Pantheon, New York.

Chomsky, Noam: 1985, *Turning the Tide*, South End.

Chomsky, Noam: 1987, 'Reinhold Niebuhr', *Grand Street*, Winter.

Chomsky, Noam: 1989a, 'Democracy in the Industrial Societies', *Z Magazine*, January.

Chomsky, Noam: 1989b, *Necessary Illusions*, South End.

Chomsky, Noam: 1991, *Deterring Democracy*, Verso.

Chomsky, Noam: forthcoming, 'Language and Interpretation', University of Pittsburgh, Series on Philosophy of Science.

Chomsky, Noam and Edward Herman: 1979, *Political Economy of Human Rights*, South End.

Davidson, D.: 1986, 'A Nice Derangement of Epitaphs', in E. LePore (ed.).

Dummett, Michael: 1986, 'Comments on Davidson and Hacking', in E. LePore (ed.).

Fossedal, G.: 1989, *The Democratic Imperative: Exporting the American Revolution*, New Republic Books.

Gardner, L.: 1987, *Safe for Democracy*, Oxford.

Goodman, Nelson: 1984, *Ways of Worldmaking*, Hackett.

Herman, Edward and Noam Chomsky: 1988, *Manufacturing Consent*, Pantheon, New York.

Hill, C.: 1975, *The World Turned Upside Down*, Penguin, Harmondsworth.

Jedlicki, W.: 1989, 'Gorbachev: An Appraisal in Human Rights Terms', *Against the Current*, May/June.

Kalven, Harry: 1988, *A Worthy Tradition*, Harper and Row, New York.

Kerstiens, Thom and Piet Nelissen: 1984, *Report on the Elections in Nicaragua, 4 November 1984*.

Kristol, Irving: 1973, *Wall Street Journal*, December 13.

LePore, E. (ed.): 1986, *Truth and Interpretation: Perspectives on the Philosophy of Donald Davidson*, Blackwell, Oxford.

Levy, Leonard: 1985, *Emergence of a Free Press*, Oxford.

Marlin, R.: 1989, 'Propaganda and the Ethics of Persuasion', *International Journal of Moral and Social Studies*, Spring.

Pastor, Robert: 1987, *Condemned to Repetition*, Princeton.

Pateman, Trevor: 1987, *Language in Mind and Language in Society*, Oxford.

Pinker, Steven and J. Mehler (eds.): 1988, *Connections and Symbols*, MIT Press, Cambridge.

Popkin, Richard: 1979, *The History of Scepticism from Erasmus to Spinoza*, California.

Preston, W. and E. Ray: 1988, 'Disinformation and Mass Deception: Democracy as a Cover Story', in R. Curry (ed.), *Freedom at Risk*, Temple.

Putnam, Hilary: 1988, *Representation and Reality*, MIT Press, Cambridge.

Rietman, Lex: 1988, *Over objectiviteit, betonrot en de pijlers van de democratie: De Westeuropese pers en het nieuws over Midden-Amerika*, M.A. thesis. Instituut voor Massa-communicatie, University of Nijmegen, Nijmegen.

Schiller, H.: 1989, *The Corporate Takeover of Public Expression*, Oxford.

HARRY M. BRACKEN

SOME REFLECTIONS ON OUR SCEPTICAL CRISIS[1]

I shall first sketch the nature of the sceptical crisis of the 16/17th centuries and then the sort of response which Descartes provides to it. Then I
shall characterize certain 20th century developments in the history of
ideas as a second sceptical crisis. It is this crisis which terrified Orwell in
the '30s. Just as the first crisis can be seen as undercutting reason to make
room for irrationality — so the second can be similarly understood as
constituted by several philosophical movements which serve to undercut
reason. Once again opportunities to create irrational structures are made
available. Descartes tries to find a way out of the first crisis. I am suggesting in this paper that Chomsky be read as trying to resolve the second.
Both do it by countering the radical scepticism with forms of founda-
tionalism — with uncovering universal features of human nature upon
which rational structures can be established.

Richard Popkin (1979) has argued that the rise of modern philosophy
is a response to a sceptical crisis engendered by the successful challenge
which the Protestant Reformation posed to European Catholic hegemony.
Calvin and Luther argue that the Catholic claim to possess the Keys to the
Kingdom is not justified. To assert that the Catholic Church is the final
arbiter of the Biblical message is to appeal to circular reasoning. That is,
the Church's assumption that it possesses the final authority to interpret
Scripture is based on its own interpretation of Scripture. Thus are generated the arguments about which group is the true Church, what are the
marks of the true Church, what are the criteria of religious truth. The
Catholic's first intellectual response, one which quickly dominates the so-
called Counter-reformation, is the wide dissemination of the arguments of
the Greek Pyrrhonian sceptic, Sextus Empiricus.

Sextus provides a detailed and extended set of arguments aimed at
releasing us from various scientific, metaphysical, logical, grammatical and
ethical 'mental cramps' so that we may achieve a healthy life of mental
quietude. (See Naess 1968.) His method is to challenge all forms of the
appearance/reality distinction by demanding a criterion for its resolution.
Some of Sextus' most ingenious arguments are directed at efforts to
establish such a criterion. Sometimes he simply demands a criterion of the
criterion ad infinitum. At other times he asks if the criterion is true. That
in turn calls for a criterion. So we appear to be caught in a circle — we
can't discern the truth without a criterion and we can't employ a criterion
unless it is true. There are discussions of such problems as locating a judge
who is not party to the dispute, of how humans differ among themselves,

59

Eric Reuland and Werner Abraham (eds), Knowledge and Language, Volume I, From
Orwell's Problem to Plato's Problem: 59—70.
© 1993 by Kluwer Academic Publishers. Printed in the Netherlands.

of the criterion for the application of logical rules — all aimed at enabling us to suspend our judgment in order to reach mental quietude.

The Church used Sextus because he provided a perfect manual for the analysis of the criterion problem. A Latin translation of the *Outlines of Pyrrhonism*,[2] by Henry Estienne, appeared in 1562. Gentian Hervet's edition of 1569 includes Estienne's translation of the *Outlines* plus his own of *Adversus mathematicos*. As Popkin notes (1979, pp. 34—35. See also pp. 68, 79), Hervet remarks in his letter of dedication to his patron, the Cardinal of Lorraine,[3] that he had found the manuscript in the Cardinal's Library and that because he thought it would prove useful in countering the Calvinists, he prepared the translation. Thus the Calvinist appeals to reason in religious matters could be neatly subverted. Religious truth could only be discerned in the light of faith — the Catholic faith.

A bit later in the 16th century Montaigne's *Apology for Raimond Sebond* gives much wider diffusion to the arguments of Sextus, since it is written in French. Montaigne spells out at length the usefulness to the Church of the criterion and other sceptical arguments. In particular, to choose another religion ought to require that one know that the new group is the true Church. But that requires a criterion, etc. Lacking the satisfactory establishment of a criterion, he appeals to 'inertia', and urges that one stay with the traditional religion of one's family and community. It is in this way that Montaigne sees Sextus as providing an intellectual barrier to changing one's religion. Of course, neither Montaigne (nor Hervet) try to prove that Catholicism is true, only that the truth-claims of the Reformers can be cast into doubt by sceptical arguments.

By the time Descartes appears on the scene towards the middle of the 17th century, both Protestants and Catholics have found it useful to employ sceptical arguments. Indeed, at least one of Descartes' teachers at La Flèche was a Counter-reformation warrior who regularly employed sceptical arguments against the Protestants. (Popkin 1979, pp. 70f.) Of course Protestants quickly learned that they too could use sceptical arguments. For example, while the Pope might be infallible, in order to recognize which person is the Pope, an infallible judgment is required. Only the Pope can make that. So the only one who can know who the Pope is, is the Pope.[4]

There were many different responses to the sceptical crisis of the 17th century. Descartes' is the most famous. Unsympathetic to sceptics and scepticism, aware that Pyrrhonism could corrode the foundations of all intellectual activity and inquiry, his *Meditations* seeks to find a secure base for knowledge. Descartes, in the course of articulating the principle *cogito ergo sum*, fully understands the threat posed by the criterion argument. That is why he claims that *cogito ergo sum* does double duty: it is at once true and at the same time it reveals the criteria marks of truth.

Descartes must go to such lengths because he is taking on a sceptical challenge which nowhere appears in the sceptical literature. The traditional Greek sceptic seeks spiritual calm, psychological ease. Descartes' famous demon deceiver — the postulated being that can cause us to be systematically misled — is not intended to generate calm. There are at least three sources for Descartes' suggestion that even the most established principles of reason can be undermined. First, there was a famous witch-trial at London in the 1630s in which, among other things, the question of the power of a possessed person to bewitch judge and jury, witness and prosecutor, was raised. (Cf. Huxley 1959; also Popkin 1979, pp. 180f.) Witch trials often devolve into formal ordeals, tests to exhibit whether one is 'possessed' precisely because ordinary appeals to evidence are thought to fail if the accused really is a witch. Second, in the late scholastic period, some theologians argue that God's omnipotence could not be circumscribed by principles of logic or by our notions of truth.[5] Third, arguments for the supremacy of the Church in all aspects of life are taken up with renewed vigor in the Reformation period. The supremacy of faith and authority over reason had long been maintained. Thus Tertullian is often said to have asserted: "I believe because it is absurd."

One complement to the demeaning of reason is the exaltation of (political) power. With a successful assault on the claims of reason, nothing stands conceptually in the way of the imposition of authority. All social and political structure finds its ultimate foundation and justification in the divine decrees mediated through Church authority. Descartes introduces the demon not as a clever philosophical game but as a step on the road to locating a truth which can not be undermined by anything. That is, a truth which is secure against any and all forces which can be arrayed against it, clerical or otherwise. If his exorcism is successful, we cannot be systematically misled by demonic forces of *any* sort. We have, within our own minds, the ultimate touchstone of objective truth.

In thinking about Descartes' efforts to locate a domain of truth which can not be tainted either by external or internal forces, it is worth recalling that Calvin is faced with a similar problem. If the [Roman] Church is indeed The Teacher, then God's message is filtered through the Church. Hence he makes the *idea* of God innate. That is not a subjectivist claim in Calvin anymore than it is in Descartes since we all have the same innate idea. By grounding this idea in the mind, rather than in empirical data, we can have access to God without any interference from those who purport to possess the Keys to the Kingdom, i.e., those who claim that without their 'instructional' guidance and discipline we cannot be saved.

The sort of rationalist option Descartes chooses is intended to provide us with a basis for the acquisition of knowledge in a context where a sceptical crisis has been created for the purpose of sweeping away all rational structures. With rationality disposed of, unconstrained authority claims can readily be advanced in all domains (including the political). I am interested in exploring how philosophical theories such as Descartes' bear on what, for want of another term, I consider political issues. Thus Cartesianism can be understood to inhibit, or to facilitate, the formulation of positions in other domains. The 17th century Church fully appreciated the risks Cartesianism posed to the doctrines they took to support their own claims to power and control.[6]

The second sceptical crisis I wish to discuss is the one George Orwell encounters in Spain in the 1930s. It is generated for him by the behavior of various groups during the Spanish Civil War. In *Homage to Catalonia* (1938), after writing about the successes of fascist propaganda he says: "This kind of thing is frightening to me, because it often gives me the feeling that the very concept of objective truth is fading out of the world." (Orwell 1966, p. 235) He adds: "I am willing to believe that history is for the most part inaccurate and biased, but what is peculiar to our own age is the abandonment of the idea that history *could* be truthfully written." (Orwell 1966, p. 236) In *1984* (1949) Orwell takes up the problem of withstanding demonic propagandistic forces.[7] This is how his Winston considers the matter:

In the end the Party would announce that two and two made five, and you would have to believe it. It was inevitable that they should make that claim sooner or later: the logic of their position demanded it . . . And what was terrifying was not that they would kill you for thinking otherwise, but that they might be right. For, after all, how do we know that two and two make four? Or that the force of gravity works? Or that the past is unchangeable? If both the past and the external world exist only in the mind, and if the mind itself is controllable — what then? . . .

The Party told you to reject the evidence of your eyes and ears. It was their final, most essential command. His heart sank as he thought of the enormous power arrayed against him, the ease with which any Party intellectual would overthrow him in debate, the subtle arguments which he would not be able to understand, much less answer. And yet he was in the right! They were wrong and he was right. The obvious, the silly and the true had got to be defended. Truisms are true, hold on to that! The solid world exists, its laws do not change. Stones are hard, water is wet, objects unsupported fall towards the earth's centre. With the feeling that he was speaking to O'Brien, and also that he was setting forth an important axiom, he wrote:

Freedom is the freedom to say that two plus two make four. If that is granted, all else follows. (Orwell 1954, pp. 67—68)

Orwell's Winston is concerned with the defense of two sorts of propositions: one whose truth depends on evidence derived from sense experience and another which is certified as true on grounds of logic. The latter, his 'important axiom', clearly has a Cartesian flavor. Faced with a sceptical crisis, faced as Descartes was, with forces seeking to dissolve the foundation of truth, Orwell falls back on a rationalist truth. He falls back on a truth that seems so secure that no challenge can put it in doubt. When we know this proposition we know something which the authorities cannot disturb. It is this radical independence of human thought which provides the basis for human freedom. It is this radical independence that reveals that we are not totally malleable — that there is a limit to the invasive power of political demonism.

But Orwell also has his doubts. What frightened him in *Homage to Catalonia* still frightens him in the passage I quoted from *1984*: "If both the past and the external world exist only in the mind, and if the mind itself is controllable — what then?" The answer is given when Winston eventually agrees that two and two make five. And yet for all that, the Cartesian dimension is clear. No where is it clearer than when he asserts that: "In the end the Party would announce that two and two make five, and you would have to believe it. It was inevitable that they should make that claim sooner or later: the logic of their position demanded it." It is a truism that our political systems are employed to control us. But it is clear to Descartes and to Orwell that nothing short of total control of our minds will satisfy.[8] And the test is always the same. People must be forced to believe what the authorities think is false. Nothing short of that suffices. It is the ultimate 'power trip'. Not only is loyalty tested, but more importantly, so is the effectiveness of their propaganda (or terror) machines. It is, then, the 'inevitable . . . logic of their position' that we must be forced to acknowledge their authority. And even a Cartesian-style reflection upon an independent and objective truth may finally succumb to the successful application of state power.

"Nazi theory," writes Orwell, "indeed specifically denies that such a thing as 'the truth' exists." (Orwell 1966, p. 236) Nazi theory was of course not articulated within the context of an explicit philosophical theory, but philosophers nevertheless were available to generate theories. Thus arguments are advanced in the 1930s and beyond which make the Nazi attack on 'the truth' philosophically plausible. This brings us to what Chomsky calls Orwell's problem: "how can we know so little, given that we have so much evidence." (Chomsky 1986, p. xxv) What are the factors that block our understanding and enhance the plausibility of, e.g., Nazism? (See Chomsky 1988b.) An important factor, of course, is propaganda. And the theories of some philosophers have helped render important intellectual attacks upon it.

Heidegger, and those who follow in his wake, apply to the traditional (as well as the Cartesian) doctrine of truth the sort of attack which had been mounted in the 16th and 17th centuries by the Counter-reformers. Just as Sextus Empiricus' assault on truth is first used by the Counter-reformers as a means for generating a sceptical crisis which would allow the Catholics to defeat the Reformers intellectually, so with Heidegger & Co. we are given a method for attacking the notion of truth. As a consequence a new sceptical crisis arises. Once again it is created in order to facilitate an expansion of political power. But notice that the philosophical position which sought to be secure against the 16th century sceptical crisis emerges as the particular target of those who were seeking to create a similar crisis in our own century. From Heidegger to our "End of Philosophy" philosophers, Descartes has been the primary target. So perhaps it should not surprise us that someone on the other side, like Orwell, should present his account of the discernment of truth in a spectacularly Cartesian form.

The sceptical crisis which is initiated in the '30s expands in the years after World War II. From Heidegger to de Man and Wittgenstein, Rorty and Derrida, the aim is to destroy both Cartesian and Kantian style epistemology. Sometimes this is done by what is now called hermeneutics — the infinite unravelling of a text, although it appears not to have been infinite for the rabbis. Beneath all the commentaries, there was for them a real text. The more unconstrained unravelling work of Fr. Richard Simon, perhaps the first post-Reformation practitioner of the art, earned him the active displeasure of his Church, and his books were placed on the Index. He had opened up a potential internal challenge to Church legitimacy that could not be countenanced. Some post World War II philosophers prefer to destroy epistemology with non-theories about philosophy of language and the nature of meaning. Such non-theories are intended to reveal that talk about any sort of independent and universally valid standpoint is a product of conceptual confusion. There are no Platonic meanings, there are only linguistic parlor games which we are trained in childhood to play and which reflect various 'forms of life'. Rhetoric finally displaces logic.[9] We even see the re-emergence of a version of the medieval two-fold doctrine of truth, this time with the 'truth' of political rhetoric replacing the 'truth' of faith — and taking precedence over scientific truth. Thus a claim can be 'true' in terms of political rhetoric yet false in terms of the facts, and vice versa. On a very generous interpretation something of this sort seems to be at work in Frits Bolkestein's (then the Dutch Defense Minister) remarks about Chomsky.[10]

In a well-known collection of essays entitled *After Philosophy: End or Transformation?* Charles Taylor (Baynes et al. 1987) says that the epistemological tradition to be 'overcome' holds that "knowledge is to be seen as correct representation of an independent reality" (p. 466). He names as four major critics of this tradition: Hegel, Heidegger, Merleau-Ponty, and Wittgenstein. He maintains that the decisive shift in focus is that from trying to locate an independent foundation of knowledge to a recognition that we are always agents in the world and hence that total 'disengagement' from it is impossible. In the same volume, Rorty, agreeing with a passage from Sartre writes:

This hard saying brings out what ties Dewey and Foucault, James and Nietzsche, together

— the sense that there is nothing deep down inside us except what we have put there ourselves, no criterion that we have not created in the course of creating a practice, no standard of rationality that is not an appeal to such a criterion, no rigorous argumentation that is not obedience to our own conventions. (p. 60)

The post-World War II philosophical scene has been increasingly dominated by movements which, for whatever reasons, converge on this point: epistemology, theories of knowledge, and in particular, the Cartesian outlook, must be displaced.[11] Our modern intellectual sceptical crisis is being created by the arguments and methods the members of these movements employ to achieve these goals. There are, as Popkin has shown, a variety of responses to the 16th/17th century sceptical crisis. We too can observe many responses. Accepting the death of truth, or at least of our capacity to recognize it were we to encounter it, various practical devices are recommended to us: for example, relativism, contextualism, and linguistic behaviorism. Even pragmatism has been resurrected, a sure sign that our suspicion that truth has been displaced by power is warranted. I shall come back in a moment to a very special sort of response.

The destruction of epistemology and the death of truth creates an open field for rhetoric and the exercise of power. Orwell saw the point. He saw that a *theoretical* movement was afoot in Spain which would contribute to making propaganda virtually irrefutable. This movement understood how to employ rhetoric, the glorification of the irrational, and the demeaning of intelligence, as foundations for constructing fascism. That, I submit, is why his Winston talks like a Cartesian — there must be some vantage point from which the falsity of the claims of power can be seen.

I have, however, left aside the 20th century equivalent of a popular 16th/17th century escape from this sceptical crisis. Finding no satisfaction in the domain of reason, some 16th/17th century figures turned to faith. Fideism became one answer to the crisis. And although the Church always claimed to be uncomfortable with fideism, and declared it a heresy, the author of a most fideistic work, Bishop Pierre-Daniel Huet,[12] prospered within the Church. For Huet, the scepticism of Sextus destroys human pretence and thus prepares one passively to receive the message of God.

Something similar occurs in our own time. When the grounds of reason are dissolved into fantasy, we are usually given new options. We too find faith recommended. Occasionally the faith is a religious one, more often as power replaces truth it has been a political cult leader. In brief, a traditional response to a sceptical crisis is the glorification of an irrational option. We do not lack examples, from the American pragmatist John Dewey[13] and his followers to Heidegger and his. We also find the 'quietist' option, i.e., those who counsel discreet support for the state by holding that an engaged philosopher thereby ceases to be a philosopher, or as

Gilbert Ryle put it at a Vienna conference in 1968, "Any philosopher who attaches himself to a political commitment ceases to be a philosopher," a political claim I think of as Ryle's Paradox.

Both Ryle[14] and Wittgenstein devote much effort to showing the more gross philosophical errors of Descartes. There is no doubt about the identity of the *primary* enemy, even if Bertrand Russell held the *secondary* enemy position in the Inferno of post World War II Anglo-American philosophy. Descartes was a target of hostility from the very outset although his philosophy proved to be attractive to certain scientists. And Calvinists, in particular, initially found his position compatible with their own. They were attracted to the idea that each individual can discover a genuinely independent truth. There are human minds and there is an independent and eternal domain which these minds can probe, explore, and know. But the real objection to Descartes is that there is no role left for those who are in the business of helping us to reach the truth. From the standpoint of (religious) bureaucracies, the trouble with innate ideas is that access to knowledge does not need to be mediated through the official teachers, the churches, the culture, or what you will. This was soon well understood. By the latter part of the 17th century, Descartes was under attack in Calvinist circles in, e.g. Franeker and Utrecht as is made clear in the excellent study by Verbeek (1988).

With that in mind I want to discuss a few aspects of the work of Noam Chomsky, aspects which have sometimes, and I think correctly, been seen as having a certain Cartesian quality. Descartes' theory is in part an answer to the question, how is it that we have the knowledge we in fact have. And his answer is at once a vigorous rejection of any attempt to ground the universality of our knowledge claims, as well as the independent status of the entities we know, in sense experience. Rejecting any and all forms of abstraction of knowledge from sense experience, Descartes turns to innate ideas. Chomsky poses a somewhat similar question in asking how a child, on the basis of a limited exposure to a language, — without any special training — develops a rich knowledge of the syntax of that language. Chomsky maintains that humans, uniquely among animals, possess a language faculty, a highly structured biological organ, whose development is triggered by a minimal encounter with linguistic data. Descartes is able to say very much less about the mind. He offers no rich explanatory theory, whereas Chomsky's account of universal grammar is a genuine scientific theory which ultimately stands or falls in accordance with scientific practice. But both Descartes and Chomsky are seeking to answer what Chomsky refers to as Plato's Problem: i.e., "the problem of explaining how we can know so much given that we have such limited evidence." (Chomsky 1986, p. xxv)

Descartes' postulation of mind is in addition driven by his concern with what Chomsky calls the 'creative aspect of language use'. By that he means "the ordinary use of language in everyday life, with its distinctive properties of novelty, freedom from control by external stimuli and inner states, coherence and appropriateness to situations, and its capacity to evoke appropriate thoughts in the listener."[15] Chomsky, like Descartes, maintains that this 'creative aspect' is found only in the human species. He also makes several speculative suggestions which are very much in accordance with the universalism Descartes espoused.[16] One deals with artistic creativity, another with moral judgment. About the latter he writes: "What its basis may be we do not know, but we can hardly doubt that it is rooted in fundamental human nature. It cannot be merely a matter of convention that we find some things to be right, others wrong."[17]

Like Descartes, Chomsky is presenting an account of human nature — an account which lies at the heart of his social and political views. It is obviously not an account which is 'deduced' from his scientific work on universal grammar and the language faculty. As in Descartes, the doctrine of human nature and established science are compatible with one another. The doctrine may suggest new directions for science as well as for ethical and political theory, but whether science can ever be expanded to handle such things as free will is extremely doubtful — a question left shrouded in mystery, perhaps because it is beyond the scope of the human intellect (see Chomsky 1975) — rather than because of an absolute 'in principle' barrier of the sort erected by Descartes. Finally, Chomsky's remarks on human nature can be read as a Cartesian-style response to the sceptical crisis of the 20th century. Thus in both Orwell and Chomsky there are efforts to exorcise a demon — a demon which now appears in the form of state power. For both Orwell and Chomsky there are two-stage responses. First, an analysis of propaganda and its role; second, an appeal to a Cartesian type universality ultimately rooted in our common natures.

Heidegger[18] is now often considered the philosophical giant of the century. It reflects a measure of intellectual honesty for a century so often dominated by totalitarians to select as 'its' philosopher a man who was committed to totalitarianism. Although the logic of Heidegger's position, and that of many of his followers, obviously does not entail that one be dedicated to totalitarianism, such steps are not surprising. In place of logic we are given rhetoric, and rhetoric is readily driven by power. In place of truth, we get 'form of life' talk, contextualism, and historicism. Note that my remarks are not directed at establishing the truth of a Cartesian or Platonic or, for that matter, any other 'objectivist' position. My remarks are only intended to expose some of the consequences of replacing, say, Cartesianism with positions which are always dialectically 'one-up' on their critics — positions which deflect criticism by the powerful device of placing putative truth claims beyond the pale. 'Revealing' the folly of objectivity, 'dismissing' truth, post-philosophy philosophy can and does make a valuable contribution to the construction of certain cultural doctrines by

rendering them impervious to rational criticism. In brief, this philosophy
can create a friendly environment for the growth of irrationalist positions.
Managers of political power know how to make good use of a vacuum —
they do not abhor one.

Heidegger and his followers correctly see Cartesianism as an inhibiting
factor when one is propagating particularism. Hence there is a good
reason for them to target Descartes for destruction. On the other hand, if
one proposes to defend universalism, and a non-vacuous account of
human nature, there is a real point in reflecting on a tradition which
retains a commitment both to logic and truth. The Cartesian side of that
tradition provides an account of the human mind as non-plastic and non-
malleable. Such an account functioned historically as a modest conceptual
barrier against racist theory, it also provided the framework within which
the first 'absolutist' doctrine of freedom of speech[19] was articulated, and
most important for our present purposes, it is a tradition in which access
is retained to a truth which is taken to be independent of historical
location, language, or culture.

NOTES

[1] Discussions with David Blitz, George di Giovanni, James McGilvray, and Eric Reuland
were helpful in preparing this paper. I am, however, especially indebted to Elly van
Gelderen.
[2] The works of Sextus are available in Greek/English in 4 volumes in the Loeb Classical
Library series. But see also Julia Annas and Jonathan Barnes, *The Modes of Scepticism:
Ancient Texts and Modern Interpretations* (Cambridge: University Press, 1985).
[3] Apparently Charles (Guise). The 1569 edition of Sextus, printed in Paris by Martinus
Juvenus, includes the section on Pyrrho from Diogenes Laertius (translated by Estienne)
plus a short piece 'against the sceptics and Pyrrhonists' by Galen (translated by Erasmus).
[4] Jean La Placette, cited by (Popkin 1979, p. 14).
[5] Gabriel Biel is sometimes mentioned in this connection.
[6] Take a different example: we can ask whether Hume's expressed philosophy facilitates or
inhibits his expressed racist views. And we can ask another (but very different) question:
Have Hume's racist views enhanced or diminished his philosophical reputation? Several
essays in (Bracken 1984) are devoted to these questions.
[7] I discuss this passage from a different point of view in chapter 6 of Bracken, 1984.
[8] See chapter 4, "Minds and Oaths," in (Bracken 1984) where I discuss several 17th
century efforts to employ oaths to control minds.
[9] There is an extremely interesting Preface, by Newton Garver, to (Derrida 1973). Garver
discusses not only the primacy of rhetoric over logic for Derrida and Wittgenstein but also
a number of other affinities and parallels between them.
[10] Their discussion, initiated with an attack by Bolkestein, contains three contributions by
each between November 1988 and January 1989 in *NRC Handelsblad*, culminating in
their 'debate' in Groningen, 22 May 1989.
[11] Chomsky has often written on the role of intellectuals in Western propaganda states.
The contributions of historians and political scientists may be more obvious, but the work
of philosophers in the support of state power already troubled Russell during World War I.
[12] See (Huet 1741). It appeared first in Latin about 1690. It was *not* placed on the Index.

[13] Dewey was an active propagandist during World War I and appreciated the role propaganda could play in the 'engineering of consent' within a so-called democracy. In this connection, see the work by Clarence Karier, cited in Chomsky, 1978 and 1979.

[14] As is well known, Ryle reviewed *Sein und Zeit* in *Mind* (1929). See also Wittgenstein, 1978.

[15]Chomsky 1988, p. 138. The creative aspect of language use is discussed in detail as early as Chomsky, 1966. Descartes' own views appear in Pt. V of his *Discourse on Method*.

[16] Descartes had little to say about political questions. But see Guenancia, 1983.

[17] Chomsky 1988a, p. 152. The topic is occasionally discussed elsewhere, e.g. in Chomsky, 1971, pp. 48—49. It also comes up in his discussion with Foucault in connection with justice and power.

[18] *Critical Inquiry*, Winter 1989, Vol. 15, no. 2, contains an interesting Symposium on Heidegger and Nazism. Contributors include Gadamer, Derrida, Habermas and Levinas. It is obviously important that we try to understand Heidegger's commitment to Nazism. One can, however, easily imagine other interesting Symposia, e.g. Aristotle and Sexism, Locke and Slavery, Hume and Racism, Mill and the Irish, Frege and anti-semitism, plus several 'major' contemporary philosophers whose involvements with various political movements or whose expressed political opinions are quite as 'distressing' as Heidegger's with Nazism.

[19] See, for example, my (1990) "Pierre Bayle and Freedom of Speech," in *Truth and Tolerance*, ed. E. J. Furcha, ARC Supplement no. 4, McGill Faculty of Religious Studies, Montreal, pp. 28—41. This is a small portion of a long-term research project.

BIBLIOGRAPHY

Baynes, K., J. Bohman and T. McCarthy (eds.): 1987, *After Philosophy: An End or Transformation?*, MIT Press, Cambridge.

Bracken, Harry M.: 1984, *Mind and Language: Essays on Descartes and Chomsky*, Foris, Dordrecht.

Bracken, Harry M.: 1990, 'Pierre Bayle and Freedom of Speech', in E. J. Furcha (ed.), *Truth and Tolerance*, ARC Supplement no. 4, McGill Faculty of Religious Studies, Montreal.

Chomsky, Noam: 1966, *Cartesian Linguistics*, Harper and Row, New York.

Chomsky, Noam: 1971, *Problems of Knowledge and Freedom*, Pantheon, New York.

Chomsky, Noam: 1975, *Reflections on Language*, Pantheon, New York.

Chomsky, Noam: 1978, *Intellectuals and the State: The Johan Huizinga Lecture for 1977*, Het Wereldvenster, Baarn.

Chomsky, Noam: 1979, *Towards a New Cold War*, Pantheon, New York.

Chomsky, Noam: 1986, *Knowledge of Language*, Praeger, New York.

Chomsky, Noam: 1988a, *Language and Problems of Knowledge: The Managua Lectures*, MIT Press, Cambridge.

Chomsky, Noam and Edward S. Herman: 1988b, *Manufacturing Consent: The Political Economy of the Mass Media*, Pantheon, New York.

Derrida, Jacques: 1973, *Speech and Phenomena*, translated by David Allison, Northwestern University Press, Evanston.

Guenancia, Pierre: 1983, *Descartes et l'ordre politique*, PUF, Paris.

Huet, Pierre-Daniel: 1741, *Traité philosophique de la foibless de l'esprit humain*, Jean Nourse, London.

Huxley, Aldous: 1959, *The Devils of Loudon*, Harper Tourchbooks, New York.

Naess, Arne: 1968, *Scepticism*, Routledge and Kegan Paul, London.

Orwell, George: 1954, *1984*, Penguin, Harmondsworth.

Orwell, George: 1966, *Homage to Catalonia*, Penguin, Harmondsworth.

70   HARRY M. BRACKEN

Popkin, Richard H.: 1979, *The History of Scepticism from Erasmus to Spinoza*, University of California Press, Berkeley.
Taylor, Charles: 1987, 'Overcoming Epistemology', in Baynes, et al.
Verbeek, Theo: 1988, *René Descartes et Martin Schoock: La querelle d'Utrecht*, edited, translated, annotated by Theo Verbeek, Les impressions nouvelles, Paris.
Wittgenstein, Ludwig: 1978, 'On Heidegger on Being and Dread', edited with Commentary by Michael Murray, in Michael Murray (ed.), *Heidegger and Modern Philosophy: Critical Essays*, Yale, New Haven, pp. 83.

THOMAS ROEPER

THE "LEAST EFFORT" PRINCIPLE IN CHILD GRAMMAR: CHOOSING A MARKED PARAMETER*

1. INTRODUCTION AND OVERVIEW

One of the largest challenges for any science is the integration of disparate sources of information. While technical proposals in linguistic theory emerge at a rapid rate, they often have have vast unarticulated implications for theory and data from language acquisition, which in turn is relevant for the formulation of the original proposals. In what follows we argue that recent arguments by Chomsky (1988) and Pollock (1989) acquire a much greater depth of explanatory power when acquisition data is incorporated. Moreover we hope that what follows has heuristic value as an example of how a close examination of acquisition data can shift, in at least a small way, our perspective on linguistic theory.

In brief, the facts of *do*-insertion articulate (1) the claim that the distinction between universal properties and language-particular features remains explicit in human grammars, and (2) inserted *do* disambiguates a parameter for the child which is unambiguous for an adult: does English have raising or lowering? Since the choices are lexically linked (*be* (and sometimes *have*) allows raising), the child is still unsure which of two options represents the regular rule and which represents the exception in his language.

We shall address these issues in light of a fairly surprising range of acquisition facts. Children produce the following forms (see more evidence below):

(1) a. "it does fits"   (tense-copying)
    b. "he did left"   (tense-copying with strong verb)
    c. "do it be colored"   (*do*-insertion with *be*)
    d. "is it is"   (copying in question formation)

Equally interesting is the fact that there are no recorded instances (to my knowledge) of:[1]

(2) a. *"it does is" or *"does it is"   (*do* with tensed *be*)
    b. *"it fits fit"   (copying with verb-raising)
    c. *"it fits do"[2] or* "it fits so" (*do*-insertion with verb raising)

The absence of cases in (2) rules out any simple performance or error theory. We will argue 1) that the appearance of extra-*do*-insertion in these contexts supports Pollock's approach, but calls for a revision in the trigger for *do*-insertion, and 2) calls for a reinterpretation of Chomsky's notion

71

Eric Reuland and Werner Abraham (eds), Knowledge and Language, Volume I, From Orwell's Problem to Plato's Problem: 71–104.

"save a derivation" to include "mark a parametric choice". In schematic terms we find *do*-insertion where there is a parametric choice (x), raising or lowering, but not where there is no parametric choice (y) (at least in terms of movement direction), only simple copying occurs (z):

(3)

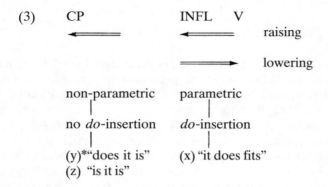

It is evident from the excluded cases in (2) that it is only the marked lowering option that triggers *do*-insertion with copying of tense. We trace now the sequence of facts and ideas that leads to these claims. Then we will sketch out the implications for acquisition of V-2 languages and two further instances, wh-copying in English and clitic-copying (doubling) in French, which suggest again that marked parametric options are explicit.

1.1. *Chomsky's Proposal*

Chomsky (1988) has made two proposals which, straightforwardly, make "surprise" predictions about language acquisition. The first proposal is a Least Effort proposal: the speaker will choose a derivation that is shortest. The second proposal is what one can call a "Default-*do*-insertion" proposal: when a derivation fails to express semantic features (contra the Full Interpretation Principle) *do*-insertion will occur to carry the semantic feature (e.g. tense). These proposals are then non-parametric constraints on possible derivations in UG. Their meta-theoretical status is unusual. They are not part of the parametric map which, putatively, defines UG and defines the steps through which acquisition must proceed. They could therefore force changes, i.e., stages, both historically and in acquisition, which are not simply progressive parametric changes.

  Children do not in fact acquire adult grammar instantaneously. It now follows that either "short derivations" or "*do*-insertion" could appear in children's grammars where they are not required in adult grammars, because certain derivations can be made in adult grammar but not in the child's grammar. Precisely this occurs, as seen above: we find that children will say "did I didn't do it", "he did left", or "do it be colored". Each of these examples involves a slightly different deviation from the adult grammar warranting, under Chomsky's hypothesis, the deviant appearance of *do*-insertion.

In what follows we will argue that the exclusion of (2) leads to a deepening of Pollock's (1989) approach in light of markedness claims: *be*-raising is unmarked and internal tense-marking is unmarked (as in *is*). *Do*-insertion in "do it be colored" then arises because the child does not know that *be* takes internal tense. The use of expressions like "he bes here" supports the claim that children may not fully grasp tense with *be*. In sum, acquisition evidence of spontaneous *do*-insertion may be the most direct and clearest evidence in favor of Chomsky's implicit claims about the meta-theoretical status of *do*-insertion, together with the additional function of marking a parametric option.

How are such mistakes eliminated from the child's grammar? The elimination of *do*-insertion in favor of an adult non-insertion analysis fits a larger theory of Defaults, which has been developed by Lebeaux (1988), as a fundamental assumption about acquisition. The default status has an important empirical consequence in terms of the subset principle (Berwick 1985). If *do*-insertion is not a default rule, then it is not clear how it is eliminated. Suppose *do*-insertion were an optionally acceptable means to express tense, as it appears to be in the grammars of children. Then it would not follow that affix-lowering would replace it: it could remain an available alternative, much like heavy-NP shift. In order for *do*-insertion to be eliminated, it must have some intrinsic characteristic which leads the child to prefer non-*do*-insertion. Lebeaux (1988) in fact argues for a whole series of defaults (in terms of Case theory, adjunction, and other cases), each of which is replaced in this fashion. Like *do*-insertion, each of these defaults predicts stages (or moments) in acquisition which deviate from adult grammar but do not reflect a parametrically-motivated difference.

The operation of *do*-insertion is one version of copying under Pollock's (1989) analysis, since properties of the verb must be copied onto *do*. Why do children prefer copies in these environments ("it does fits") where adults prefer traces? Such questions are usually addressed implicitly or explicitly in terms of "performance".[3] It is "easier" for an adult to say nothing and therefore a trace is sufficient. It is "easier" for a child to have a copy which indicates a D-structure origin. Yet it is hard to understand why performance demands would be the opposite for child and adult. It seems far more natural to seek a difference in grammars which underlies the putative performance difference.

To answer these questions we follow the spirit of Chomsky's proposals in a different direction. We argue that: children put a copy in a structure whose parametric status is underdetermined. We will argue that the existence of a copy for a child where an adult would have a trace has nothing to do with "performance", but rather it is an instance of a child marking one of two parametric options when the parameter is not yet fixed. In particular, copying arises when children have not fully determined whether verb-raising or affix-lowering is the unmarked case in their language. Once again, different lexical items are linked to raising (*be* or *have*) and lowering (*-ed* and *-ing*). Therefore the child receives information which supports both sides of the parameter. Consequently the decision cannot be immediate. This is not such a radical departure from the usual conception of a child's grammar as a consistent synchronic object.[4] The deepest scientific principles, like gravity, are always imperfectly expressed in the real world. The logic of this approach suggests that adult grammars may also be, in part, parametrically unresolved, particularly when they exhibit copying. Chomsky (1986, p. 17) has suggested something similar:

... a speech community of uniform speakers each of whom speaks a mixture of Russian and French (say an idealized version of the 19th century Russian aristocracy). The language of such a community ... would not represent a single set of choices among the options permitted by UG but would include "contradictory" choices for certain of those options

Adults build up such sublanguages in morphology where Greek and Latin words allow different affixes (*civility* but not **evility* (J. Randall, personal communication). The same phenomenon could be at work in building up verb classes which take inconsistent options in UG.

This view, in turn, fits suggestions by Chomsky (1984) and Pesetsky (1989) to the effect that: the distinctions between universal features and language particular features are retained in adult grammar. This means that the adult knows which features of a rule are universal and which are language particular. It also leads to the assumption that the adult retains some knowledge of unchosen parametric options. We turn now to a broader discussion of the parametric model and then a more precise presentation of the argument just given.

1.2. *Acquisition Theory and Developmental Evidence*

It is important to articulate the fact that our goal is to provide a theory of how acquisition may occur, not a development sequence in which we state exactly what a child's grammar is at a certain stage or age. Knowing what a child's grammar is at a particular age is of interest, but it is no more necessary than knowing what the grammar of one individual is in writing a grammar of English. There are no full grammars of any individual speakers of Standard English. We do not require of theoretical work on intuitions that it guarantee that the grammar of any one individual be explicit or even consistent. Nor should we require of acquisition theory that any putative "stage" be completely explicit or consistent. Because our goal is to articulate a theory, we will use evidence that is fairly rare and drawn from different children at different ages. We treat them as reflections of logically necessary points in acquisition which may have been crucial at an earlier point in acquisition.

The fact that the evidence may be "rare" might lead one to believe that it is marginal, or as is often said, a reflection of "performance". A child may pass through stages silently. It could easily be the case that only 10% of the decisions that a child makes leave overt evidence. For instance, we do not expect to see the child choose the Head-parameter, since input may allow an instant choice (VO or OV). Therefore the naturalistic data is used only as a source of clues through which to develop a model that meets the primary question: how is acquisition possible under any assumptions? It simply shows that once we obtain a possible theory, we can then seek to answer questions about which features of the acquisition process are explicit and which ones are implicit, which remain in the grammar as lexical exceptions, which disappear slowly and which disappear instantly.

Naturalistic data and experimental evidence may also, at the stage examined, understate a child's grammatical knowledge and nevertheless provide insight into what earlier stages must have been. Demonstrations that a child at a given age does or does not know some feature of grammar simply fail to address the logical problem of acquisition.[5] The child's variation at a given age, unless the same variation holds for an adult, suggests that the child has more than one dialect, one of which is an earlier stage of acquisition.

There are, now, a number of proposals in acquisition theory which address the Primary Linguistic Data problem, i.e., they structure the fashion in which input is allowed into the system.[6] Lebeaux argues for a variant of the lexical learning hypothesis:[7] children attack an input by projecting either 1) a D-structure based on lexical content (theta-structure) or 2) an S-structure based on a surface string, for which the D-structure is not completely evident. In brief, two predictions which follows from Lebeaux's analysis are that:[8]

(4)   All movement sites can be directly generated.

(5)   S-structure has two possible D-structures.

We will call these monolevel representations because though both

D-structure and S-structure are conceptually available to the child, the connection for a given structure may be opaque. There is direct acquisition evidence from Davis (1987):[9]

(6)   "are you put this on me"
        "are you get this down"
        "are you help me"
        "are you know Lucy's name is"
        "are you want one"
        "are you got some orange juice"
        "are this is broke"
        "are you don't know Sharon's name is"
        "are you sneezed"

The fronted auxiliary is initially analyzed as an independent question morpheme which seems to be directly generated without an IP origin.[10]

The assumption (5), that the learner can project an S-structure without being certain of the corresponding D-structure, fits the possibility that there is parametric indeterminacy at certain stages of acquisition. We will explore the consequences of (5) for *do*-insertion and then briefly, for wh-movement in what follows.

1.3.   *The Lexical Representation of Parametric Knowledge*

How is a child's knowledge represented in an indeterminate phase? We argue that the potential for parametric ambiguity is a natural corollary of the Lexical Learning hypothesis,[11] in the following way. Subcategorization frames allow the existence of, in effect, complex lexical items. If subcategorization frames can be individually represented, then they can have unique and even idiomatic structure (DiSciullo and Williams, 1987):

(7) a.   a good time was had by all
    b.   *all had a good time

The idiom exists only in the passive and the bare quantifier is not a possible subject in modern English.

Now if children build up sets of lexical items, including their sub-categorizations, then the subcategorizations can contain the output of transformations (under the hypotheses above) or parametric settings. In addition, one set of items could have one parametric setting and another could have a different one. The child, for a period of time, might remain unsure which parametric setting is productive and which is exceptional. Let us illustrate.

In English, there is an exceptional use of matrix clause *seem* with pro-drop. A National Geographic article begins with (8) which is not possible with (9):

(8)   Seemed pretty as a picture.

(9)   *appeared pretty as a picture

The child could then build up two sets of verbs: one set allows pro-drop (*seem*), and one set does not (*appear*). The parametric decision could still depend upon a different syntactic analysis: recognition of expletives. The presence of expletives then sets the parameter against pro-drop and marks one verb set as lexical exceptions. This model would allow the child to countenance contradictory evidence without necessarily changing the grammar back and forth or ignoring all evidence that did not fit the current parametric hypothesis. In fact many transformations remain linked to a class of lexical exceptions. For instances, *tough*-movement is possible with only a limited class of adjectives. The grammar of verb-movement has just these characteristics. Pollock argues that inflection lowers in English, with the exception of *be* and *have* (see below). Thus the child is confronted with neutral data (*John runs*), data which favors verb-move-ment (*John is always happy*), and data which favors lowering: (*John always sings songs*).[12] The prima facie facts invite, therefore, just the kind of model we are advancing.

2. POLLOCK'S APPROACH TO IP

We begin with a brief look at IP structure. Pollock (1989) has proposed an elaborated IP in which several elements which have often been repre-sented as Heads or only affixes are represented as full Maximal Projec-tions with the power to function as barriers and block the assignment of a theta-role. The claim crucially explains why *do* appears with negation, but not in simple declarative sentences:

(10) a.   John hits the ball.
     b.   John does not hit the ball.

A separate Tense-Phrase and Modal Phrase precede a Negative Phrase. In English the tense-marker will lower to give the verb tense unless a nega-tive intervenes because the negative blocks this movement. *Do*-insertion then occurs to carry the tense which preserves adjacency between the verb and its complements (10b). The structure involved looks (roughly) like this:

(11)

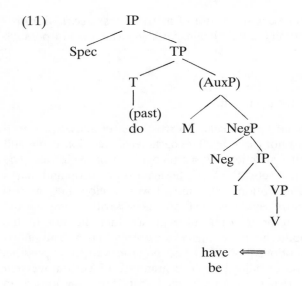

The *do* appears if the tense is stranded: TENSE must attach to a verb which is a Head.

The *do*-insertion rule, under this approach is a "substitute verb" that is required only when the verbs in question must assign a theta role and which copies the theta properties of the verb. Where the verbs do not assign a theta-role, they can freely raise over negation. This arises in English for verbs like *be* and *have*. Pollock notes these facts and offers an explanation:

(12) a. *John doesn't be happy. (John isn't happy)
     b. *John does not have gone. (John hasn't gone)
     c. *John did not be singing. (John wasn't singing)

"It is tempting to assume that those facts follows from the correct definition of what a substitute verb is. . . . Earlier we analyzed auxiliary *be* and *have* as verbs whose lexical entries lack a theta-grid. Thus it is plausible to assume that in (11) (the trace of) *do* does not have anything to copy, thereby remaining semantically empty." Therefore ungrammaticality results in (12).

The phenomenon of *do*-insertion is a special feature of English, different from French, because in English the tense marker lowers while in French the verb raises. The consequence is that in French an adverb can appear between the (raised) verb and the direct object, while in English nothing comes between the verb and the direct object:

(13) a. he always opens doors
     b. *he opens always doors

    c. Pierre lit toujours de livres (he reads always books)
    d. *Pierre toujours lit de livres

How does a child fix these subtle features of grammar? In general adverbs have great freedom: (*always*) *John* (*always*) *can* (*always*) *play* (*always) *ball* (*always*). Why would a child assume that there is just one position that is ruled out? The first question to ask is when children attain this knowledge, unless UG plays a role?

    As a background for this question, let us observe again that the child receives inconsistent evidence.

(14) a. John runs
     b. John is always happy
     c. Bill always sings songs

In (14a) both raising or lowering could have occurred. In (14b) raising must have occurred, while in (14c) lowering must have occurred. A tightly constrained linguistic system should allow the child to correctly analyse each of these sentences instantly. However, it may be unclear which is the lexical exception and which is the general rule.

    The language which the child hears is in fact more complex:

(15) a. be careful
     b. Don't be late
     c. "the Hostages are in Lebanon, but we know not where" (Ted Koppel)
     d. It matters not what you say

It is evident from (15b) that *do*-insertion can occur with *be* in imperatives, and (15c, d) reveal that the raising operation over *not* is retained by adults, but limited to contexts of special gravity. How does the child know that these uses are exceptional?

    One might assume that the lexical class of *have* and *be* is defined in UG as undergoing raising. Then no learning would be necessary. However there is evidence that the child does not locate the bounds of this exceptional class immediately. One child consistently said:[13]

(16)   "what means that"
       "what calls that"

The question is formed from the IP and entails verb-raising, though the child certainly never had any adult examples of this kind. Therefore, it must be a principled overgeneralization. It is noteworthy that both *call* and *mean* are roughly in the equative class of verbs. If the UG generalization is that equatives always raise, then the child must learn for English that the generalization is even more limited.[14]

2.1. *Acquisition Facts*

In computer searches I examined hundreds of cases of *do*-insertion,[15] and specifically searched for combinations of "do + be", "do + is", "do + was".[16] In addition I performed a search of all "-ly" structures in Adam and then used a set of common adverbs to perform searches on the other children in the Childes corpus. The adverbial phrases searched for include: "maybe", "still", "even", "really", "only", "probably", "always", "never", "sometimes". I present here representative data taken from these children. A more careful search of this data with other adverbs, larger contextual windows, and careful age correlations would be in order. The purpose of these searches was to establish the existence of a set of phenomena. Exactly how each child progresses through them remains an important research topic.

There is straightforward evidence that children are not raising main verbs to the tense position, because:

(17)   There are no instances of an adverb between the verb and direct object.

We find (18a), but not (18b):

(18) a. "he always close doors"
     b. *he closes always doors"

This is the opposite of French,[17] nonetheless, the analysis of raising seems to be in place as soon as adverbs and auxiliaries are available in English because in my searches of several thousand adverbs, none like (14b) ever occurred. The pre-sentential position, post-sentential, post-subject, and post-auxiliary positions are all used:[18]

(19)   "we always do that at school sometimes" (Adam)
         "Even I want you to drive me to school" (Tim)
         "I once did it last night" (= I did it once last night) (Tim)
         "Do you know what even was happening" (Tim)
         "why does sometimes Andy doesn't look like Andy"? (Abe)
         "Daddy doesn't mostly get it" (Daniel)

I have chosen unusual examples to reveal the productive use of these adverb positions (most of them have scope ambiguities that would make them unacceptable to an adult). None, however, appear between the verb and the object. The post-verbal position for adverbs, predictably, occurs only with the verb be: "Laurie is always a Mommy too".

2.2. *Adverb Barriers*

There are a few very precise examples of sentences in close sequence which reveal that the child operates with an articulated IP structure where an adverbial node can block tense-lowering. Consider:

(20) a. "Is that my meat/It maybe be my meat"
     b. "It maybe be dark . . . it maybe be dark"

(21) a. "I always be a mummy"
     b. "Laurie is always a Mummy too"

(22) a. *Laurie be always a Mummy too"
     b. *It be maybe mine"
     c. *be it dark?

The tensed element occurs before the *always* form (20b, 21b), but not after (20a, 21a). These examples point again to the articulated IP structure where an intervening node must be preventing a connection between *Tense* and *Be*. In these two examples we see that the *Tense* does not lower onto the verb just where there is an intervening adverbial element: *maybe* and *always*.

(23)

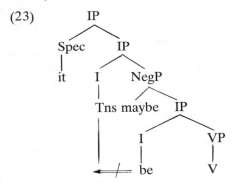

This suggests that adverbs can play a barrier role in the transmission of tense as well as negation. This leads to the hypothesis that the NegP node which we have utilized should perhaps have a broader definition: AdvP. The one case maybe in fact seems to be both negative and adverbial.[19] Other examples from Adam:

(24)   "Robin always be naughty"
        "because Indians always be bad"
        "he always get to nursery school"

It is clear that the structure prevents the verb from connecting to the preadverbial tense position.[20] Despite the fact that the IP domain is one where considerable language variation exists, it is has been precisely triggered.

3. *DO*-INSERTION

Under the predictions above, we would expect that *do*-insertion, being an in situ default representation of structure, should be immediately available to children.[21] Adam, like all the others, shows *do*-insertion in the adult manner (40 = protocol number):

(25)   40 "you didn't change it, did you"
         41 "I didn't see no tigers"
         42 "I didn't put no pant on"

This is not insignificant. Were *do*-insertion to be the last feature of grammar to emerge, we could readily explain it as a marginal phenomenon which one would expect to be difficult to acquire. Instead it is a default rule that a child has immediate access to in UG.

We turn now to a variety of other contexts in which *do*-insertion occurs (first extensively analyzed by Mayer, Erreich, and Valian (1978). These cases do not usually exist in isolation, which is why there has been a strong tendency to regard them as performance "errors". But some forms never occur. This cannot be predicted by performance alone.

The existence of the following forms in acquisition is perhaps the best straightforward evidence that the child attempts to project tense as an independent node:

(26) a. Strong Verb cases [from Davis summary]
         "I did broke it"
         "I did fell when I got blood"
         "What did you bought"
         "Jenny did left with Daddy"
         "I did rode my bike"
         "he could caught that"
         "What did you found"
         "What did I told"

         [from Roeper]
         "did you broke that port"
         "why did you left your extra keys at home"
         "Did I ate here or did I ate here"[22]

     b. Regular Verb cases [from Davis]
         "I did fixed it"
         "the plant didn't cried"
         "they didn't spilled"   "she didn't goed"
         "why I did break it"   "I didn't missed it"

   c. Present Tense cases [from Davis]
      "does it rolls"   "they don't likes to fly"
      "does he makes it"   "does it opens"
      "why doesn't this goes off"
      "why doesn't we has a marble table here"

   d. Non-copying *do*-insertion [Roeper corpus, in 16 days]
      "I did catch that bee"   "I did jump in"
      "I did put it on"   "I did jump"
      "I did turn it off"
      "I did scare a kitty away"
      "did fall down"   "did find a butterfly"
      "who did take this off"
      "I did paint this one and I did paint this one and I did paint this
      one"

Pollock argues that *Tense* is an Operator and therefore prefers a two-position representation. Once again, we find that a monolevel representation in which both positions are articulated at S-structure is evident: tense is marked both on *do* and on the verb.[23] This can be construed as direct confirmation of the Least Effort principle in acquisition.

3.1. Do-*insertion in* Be *Contexts*

In addition, we find it occurs precisely where Pollock rules it out, in *have* and *be* environments:

(27)   *Do*-insertion for *be* (from Davis: Brown, Cromer, Pinker, Roeper, Valian, deVilliers, personal communication)
       "you don't be quiet."   "didn't be mad"
       "this didn't be colored"
       "did there be some"
       "does it be on every day . . ."
       "does the fire be on every day"
       "do clowns be a boy or a girl"

(28)   *Have*
       "it's don't have any oil"
       "it didn't has any"
       "this don't had a nap"

Pollock's criterion is incorrect: *be*-forms can elicit *do*-insertion. What then is the crucial factor?
   L. Green (1990) observes that Black English exhibits precisely the

same kind of *do*-insertion with *be*: *do he be sleeping*. Now the question arises: what allows *be*-raising in BE and what prevents it in child language. One could argue:

(29 a.   that the "be" form does assign a theta-role and therefore is identical to other non-raising verbs, or

b.   another feature of meaning, namely Aspect, prevents raising, or

c.   that some feature of the agreement system is involved.

Green argues, based on aspectual characteristics of BE, that aspectual verbs do not raise.[24]

One can also argue that richness of agreement may be involved. In child language and Black English (BE), there is evidence for inflectional agreement, but in both systems the children do not immediately get the full paradigm (*am, are, is*). In BE there is a preference for "I is", "you is", "he is" and in child language one often finds the same (or "I are . . ." or "he be's here" and apparent random variation). It is true that aspect varies in child language as well "sometimes I be dry in the morning" is found for the habitual reading, and for what one might call the "generic fantasy" common among children we find "you be the Mommy".[25] The latter "subjunctive" reading holds for adults as well. It is clear that some subtle features of aspect are not initially controlled by children, while other aspectual distinctions appear quickly.

Nonetheless, the view that *be* raises only when there is a full analysis of the paradigm supports the suggestion by Jaeggli and Hyams (1987) that the "morphological uniformity" of the verbal paradigm is crucial. If the verbal paradigm involves no endings, or required endings, then it is uniform. If the paradigm is mixed, some endings are present and some stems are present, then it is non-uniform. Until the *be*-paradigm is securely analyzed as non-uniform, children will not allow *be* to raise.[26] This obviously calls for a deeper explanation. We have the structure (a):

(30) $[_{IP}[_{spec}$ he] $[_{IP}$ is$_i[_{NegP}$ not $[_{VP}$ t$_i$ here]]]]

Some strong feature of agreement (i.e., paradigmatic differentiation) is needed to make the connection between the verb and its trace. In effect, then person and number agreement, if relevant, must be matched in order for raising to occur. By hypothesis, when the strong agreement system emerges in child language, the *do*-insertion option is dropped. Until that point, both the original verb position and the tense position are directly represented in a monolevel analysis.

This then represents another form of lexical constraint on syntactic systems. In effect, the system remains lexical if the paradigm is incomplete. We can state the phenomenon in this fashion:

(31)   Incomplete paradigms cannot support syntactic generalizations.[27]

Syntactic generalization then has the form of substituting a category for a particular verb:

(32)   *be* ⇒ raising
             generalizes: V ⇒ raising

(33)   *push* ⇒ attracts tense lowering ⇒ push + ed
             generalizes:
                 V ⇒ attracts tense lowering ⇒ V + ed

Incomplete inflectional paradigms cannot undergo generalization. This line of reasoning seems to have promise, but a more refined analysis shows that raising is also affected by markedness considerations which we now discuss.

3.2. *Lexical* Do-*insertion*

Pinker (1984) makes an important observation about the contexts of *do*-insertion in child language: it predominantly involves strong verbs. We find strong verbs under two conditions: tense-copying ("did broke") and non-copying ("did break"). In addition to the cases cited above we find:

(34)   "I did broke it"
             "Jenni did left with Daddy"
             "you did hurt me"

and *do*-insertion instead of tense-copying:

(35)   "what you did eat"
             "I did see it"

These are not cases of free variation. A few exceptions exist (like "this didn't has any") but otherwise we have relatively few reports of cases of tense-mismatch like:

(36)   *"he does left"
             *"he did comes"

There could still be a stage where tense is not figured out lexically or misanalyzed, as some have claimed, but the instances of tense-matching are far too numerous for one to claim that there is random variation. Note that if there is a stage where the lexical identity of different affixes is unclear, this would be quite different from the assertion that no tense knowledge is present.

What then is different about the strong verb system? Both the verbs which permit raising allow internal tense marking (*have* and *be*). Therefore the child may be following a markedness system which reflects these preferences. Markedness system:

(37)   Raising is unmarked

(38)   within the marked lowering system:
        lowering of an affix (-ed) is unmarked (push + ed), lowering a
        semantic marker is marked (past).

(39)   internal tense-marking requires lowering of a semantic feature,
        which is marked   (leave + past ⇒ left)

Therefore, a child prefers to raise internally marked verbs, and lower
affixes. Any input that is at odds with this is marked. Consider these cases:

(49) a. John does sings ⇒ marked because lowering occurs
     b. John did left ⇒ marked for lowering and lowering a semantic
         marker

Therefore we have a correlation with the fact that the strong verbs are
more likely to exhibit copying and hence *do*-insertion if lowering is
necessary. And strong verbs are more likely to raise if we assume that
affixes are, in a sense, designed to move (in this case to lower). This
suggests again that *do*-insertion, at a more subtle level, registers derivations that are marked in terms of UG.

   In fact, adults will also prefer (41a) to (41b) if given a choice:

(41) a. ??it does fits
     b. **it does is

This shows that adults register the unmarked nature of raising. In effect,
then, there is an unmarked form of lexical substitution in UG: substitution
on a complex element [verb + past], rather than on the element [verb +
ed]:

(42)   UG substitution: verb + past ⇒ was (unmarked)
                          verb + ed ⇒ was (marked)

This argument has interesting implications for the notion that lexical
insertion can occur at different points in the grammar. The concept of
parallel morphology, which allows lexical insertion at different points in a
derivation, has been pursued by Hagit Borer (forthcoming).[28]

3.3. *The Excluded Form:* Do-*insertion and Tensed* Be

Now let us ask again, exactly why the child would say "Jenni did left"? The
answer is that the child is exposed to a contradiction: Internally marked
tense should prefer raising. But the presence of preverbal adverbs in adult
input ("John has always left") means that the grammar prefers lowering.
The form *Jenni left* is ambiguous while the form *Jenni did leave* indicates
that the grammar has chosen lowering and that the unmarked preference

for *left* (i.e., raising) is not chosen. In other words, the copying indicates the parametric choice. The *do* is inserted infrequently with regular verbs because for regular verbs the lowering analysis agrees with the unmarked case for lowering, affix-movement, although even in this case, the *do*-insertion continues to mark the parametric choice.

This leads to a crucial prediction: *do*-insertion will not occur for the completely unmarked case (43a, b):

(43) a. *"John does is here"
     b. *"does it is here"
     c. "does it fits"
     d. "do it be"

We have examined over 200 examples of auxiliary errors (drawn from the appendices of Davis (1987), as well as from our own data) and found no examples like (43a, b), with the exception of a few fixed forms.[29] This prediction follows because when the child understands the tense on *is*[30] then it has found the unmarked case for raising. Raising is preferred for strong verbs including *be*, so two forms of unmarked case match. *Do*-insertion occurs only when a marked parametric option is chosen. We then predict the presence of (43c), which does occur. We can also predict that (43d) occurs. Children often attempt to regularize be and say "he bes here" as if it had no internal tense form. It remains to be shown that these two forms correlate in the grammar of particular children, and possibly that they correlate with an aspectual usage of *be*.

Recall now that strong verbs ("I did broke it") are the most frequent locale for *do*-insertion. Therefore it is not the fact that *is* has internal tense alone that leads to the absence of *do*-insertion, it is the fact that *be* raises while other strong verbs do not raise. The evidence for the child, once again, is that adults do not say *I broke always the door* and consequently neither do children. Further evidence comes from *be*-copying.

3.4. *Copying and* Be: *The Non-parametric Piece of the Chain*

In a sentence of the form *is John here* there is a chain with two traces. There is raising from the VP into a Tense Node (part of IP) and then raising into CP (where TP, following Pollock, is like IP):

(44) $[_{CP} [_C \text{is}_i [_{TP} [_{T'} \text{John} [_{TP'} [_T t_i [_{VP} [_V t_i]]]]]]]]$

                ⟵——— ⟵——— raising

                      ———➤ lowering

        non-parametric   parametric

The two pieces of this chain have different parametric status. One part of it is directly subject to parametric raising/lowering variation while the other is not. Therefore we predict that *do*-insertion can occur with respect to one part and not the other. The *do*-insertion provides evidence of the origin of the chain with respect to lowering.

Although the SAI operation, with respect to chains, is very much similar to lowering, it does not have the same parametric status. The inversion operation may be connected to the parametric system in a different way: some languages do not signal question-formation with inversion. Within the above derivation, there is only one parametric ambiguity and therefore only one position where *do* can be inserted. There is direct evidence on behalf of this view.

The effect of *do*-insertion is to create a copying environment in many (though not all) instances. Suppose copying alone were the basis of *do*-insertion, then we would predict that *do*-insertion would occur wherever copying of the tense marker occurs. Therefore it would occur in both portions of the chain we have outlined above. In fact, it never occurs with respect to *is* although as has been widely reported that copying does occur with the verb *is* (see Davis 1987):

(45)   "what's he's doing"   "what's the mouse is doing"
        "why is there's big tears"   "what is the woman is doing"
        "Is Tom is busy"   "Is it's Stan's radio"
        "Is this is the powder"   "Is that's a belt"

Again, we have found no examples of the form **what does the mouse is playing*, although as pure performance errors one might expect to find at least one or two (i.e., the child says "does" meaning "is"). Once the verb has raised to *Tense*, then there is no difficulty in moving to the pre-subject position.

The reader may have observed that many of our examples involve inverted *do*. In each instance, though, one must argue that the *do* is first inserted in IP and then inverted. Or, predictably, once inserted, it can be copied:

(46)   "why do deze don't unrase"
        "why did you didn't want to go"

Note again, if only the semantic past tense marker were inverted, then we would predict the presence of *"did you was here" at the stage where children say "did there be some".

The notion that children identify tense before they do SAI entails a further prediction: no inversion without *Tense*. There are no reported examples of question formation without *Tense*.

(47)   *"what be that"

Only "what is that" and "what is he doing" occur. There are thousands of questions that begin with "is NP", but no one has reported a question of the form *"be NP".[31] Under the common hypothesis that these inflectional forms are in "free variation" for a period of time, the absence of these forms is surprising. These fact supports the view that there is only one way in which to form questions: by movement of a tensed element into the COMP position.

(48) $[_{comp}$ is$_i$ $[_{IP}$ NP $[_{TP}$ t$_i$ $[_{VP}$ V]]]]

This must be a feature of UG which belongs to the unmarked core of grammar.

3.5. *Summary*

Assume that both (a) Verb-raising and (b) unitary lexical insertion are unmarked, then we make the following predictions, under the assumption that *do*-insertion occurs only when some part of the analysis is marked.

(49)   "It is" ⇒ *unmarked*, no *"it does is"
           "it does fits" ⇒ *marked* because lowering is involved
           "he did ate" ⇒ *marked* because lowering involved
           "he do be sleeping" ⇒ *marked* because no internal tense, so raising is not the unmarked case

In sum, all of the forms that involve *do*-insertion are demonstrably marked in some form. The core of these examples is the marked character of lowering as opposed to raising.[32]

3.6. *Individual Variation*

There is an important limitation in the acquisition data here. We do not have a microscopic account of how individuals develop. It is possible that children actually move through stages where they have "left", "lefted", "did lefted", "did paint", "painted". Or it is possible that different children manifest different variants, or that the variants are co-temporaneous. It is not possible yet to see if a micro-evolution, which could occur in a matter of days, does occur. There are some examples which suggest that individual children actually continue to be aware of the different alternatives. Pinker (1984) cites these cases from Erreich, Valian, and Winzemer (1980), produced in close sequence by a child:

(50) a. "where goes the wheel"
     b. "where the wheel do go"
     c. "where does the wheel goes"

The predictable unmarked form occurs first: *goes* has internal tense and therefore undergoes V-2 raising. The verb "goes" here is not surprising, since it has closely related constructions like "here goes the wheel" where

V-2 does appear (compare: *Here ran the man*). This variation is not surprising if we assume, as Chomsky (1988) suggests (in reference to second language work by Flynn (1987)), that the mature speaker retains the distinction between unmarked UG phenomena and language particular decisions. It indicates that both the verb-raising option and lowering are available, but the child is uncertain about whether it applies to the verb in question.

Since verb-raising is, putatively, unmarked we can make the prediction that the direction of the child's self-correction would always go from verb-raising to verb-lowering, as it does above, and never the reverse. If the grammatical shift were some form of pure "performance" errors, then we would expect the variation to be random. Unfortunately we do not have a child corpus with sufficient refinement to verify this prediction numerically.

Other variations occur, all of which fit the mode. Some children inflect *be* and produce *bes*, just as the child who says "does it be on every day" appears to have not yet identified the tense marked for *be*. We predict, but have not verified, that the same child does both.[33]

There is a counter-argument to our syntactic claims about the role of *do*-support: sentences with an extra auxiliary have a different pattern of intonation, emphasis, and even truth value. Could the presence of extra *do*-insertion be motivated in terms of illocutionary force? Consider the following conversation (which I recently overheard) from two 4 years olds:

(51) a. "I don't want to go outside"

and then

b. "Do you don't want to go outside".

In the narrowest sense, one child wants to know if the other child's attitude agrees with his. Parallel syntax equals parallel attitude. The child has achieved a literary effect. It is one which his grammar currently allows, but which will be eliminated when the *do*-copying option is eliminated, because the sentence "do you not want to go outside" has a slightly different force.[34]

This reflects on the modular character of acquisition.[35] Each module undergoes a partially independent set of changes. The External-language (in Chomsky's sense) produced by the interaction of modules is a diverse and complicated product. Consequently possible unambiguous readings may be lost in the growth of grammar.[36] The syntactic module may be simplified at the cost of a semantic distinction. This indicates that simplicity is not measured with respect to the whole grammar, because it may be that a simple syntax/semantics relation is lost in this process of syntactic simplification.[37]

While one might suppose that the reason that *do*-insertion exists is to preserve such options, we can see that it would prevent numerous changes in the system. We conclude that subtle semantic variation cannot explain the presence of copying, but rather reveals the modular limitations on the grammar.

4. *DO*-INSERTION SUMMARY

We have presented a special angle on *do*-insertion copying phenomena in acquisition: we argue that it provides an overt, monolevel, representation of a derivation which, without the explicit copy, would be systematically ambiguous. The ambiguity may exist in the child language, but not in the adult language, because it reflects different parametric options, not just different derivational options. There is a parametric ambiguity between raising and lowering in the grammar which *do*-insertion resolves. In the adult language, there is only lowering as a regular rule and therefore the ambiguity does not exist. This analysis provides independent support for the view advanced by Lebeaux that acquisition data reveals default options, and for the view advanced by Chomsky that *do*-insertion is intrinsically a default option.

Two subtle facts are predicted if this approach is pursued: 1) the presence of *do*-insertion copying in lowering environments ("it does fits"), and 2) its absence in raising environments with *be*. Does the same argument hold elsewhere?

5. COMP: THE LANDING SITE FOR *TENSE*

The claim that English copying behavior reflects subtle features of grammar is underscored when we consider the variations found in German. Evidence from German (Weissenborn 1988, Weissenborn and Verrips 1990, Weverink 1989, Pierce 1989) supports the claim that children differentiate verbal positions (VP, IP, CP). In particular, the Germanic evidence suggests that children distinguish finite and infinitival forms. The crucial fact which emerges from these works, as these works show, is that there is a complementary distribution: no infinitives appear prenominally in Dutch and German. For example in Dutch, from 1.10 yr child:

(52) a. doe je (do + fin you)
zie je (see + fin you)

b. limonade drinken (lemonade drink + inf)

c. "papa schoenen wassen"
[Daddy shoes wash-infinitive]

d. *"wassen papa schoenen"

Contrary to English, there is no stage at which children are limited to a single OV or VO order. Why should there be variation in word-order in child language in Dutch or German but not in English?

This evidence suggests not only knowledge of a *Tense* node, but knowledge of a transformation from the earliest stages. It suggests that the capability to project transformations is immediately evident if the triggering information is not opaque.

Now we must ask what kind of transformation is involved. There is currently a debate about whether the underlying structure is (53x) or (53y):

(53)   x)   I O V   or   y)   C O V I

For the stage where there are questions, we assume the presence of CP, but it is still unclear if there is a medial or final IP.

(54)   C . . . I O V   or   C . . . O V I

The early stage appears to be, superficially, ambiguous on this point, because we do not know if movement goes to the I or the C. Other Scandinavian languages are like English in having the I in the middle.

A look at the copying data provides an interesting point of entrance. If the German child has, at the beginning of the sentence:

(50)   C IP = Spec C Spec I
        I can ⇒ Can I can

then the structure is identical to English and we would expect a copy form, found in hundreds of children:

> "can I can"
> "Is John is busy"

Although the German data is not as extensive as the English data, it is substantial. We find no copying in cases like (56)

(56)   *"kann er kann den Ball werfen"

In fact, copying structures do occur in German. These are reported by Tracy (1990), Meisel (1990), and Meisel and Müller (1991):

(57)   "wo is das Junge is" (where is the boy is)
        "wir hamen der Mond un das Rotkappchen und ein Zwerg geshen hame (= haben)"
        (we have the moon and Rotcappchen and a dwarf seen have)
        "die ham das schön gemach hatten" (they have that already done had)
        "Jetzt sagt er das sagt" (Now says he that says)

It is the final position which receives the copy, which suggests that the I is in final position. It is not the initial verb position which is marked, but rather the verb moved to the Tense position. This suggests that the children do not have a medial I at this point.

The I is precisely the same position which is marked in English when children produce "can I can come": the tense position after raising of a modal:

(58) $[_{CP} [haben_i \ldots [_{VP} gemach [_I Modal t_i] [_T hatten_i]]]]$

$[_{CP} can_i [_{IP} [_{spec} I] [_{TP} [_T can_i + tense] [_{MP} t_i [_{VP} [_V come]]]]]]$

Why is it the tense position, and not the original verb position which receives the copy? In fact, there is a three-part chain in both English and German. The origin (VP) and the landing position (CP) are clear in both languages. It is the middle element, which is marked by the child in both English and German. It is also the middle position which is subject to a great variety of parametric factors and which has no thematic projections.

In effect then each language, English and German, exhibits copying phenomenon with respect to the verb chain. If the hypothesis is correct (see Chomsky 1986, Rizzi, to appear) that the subject-verb inversion in English is a "construction-specific" rule and hence marked, the English children are indicating that some feature of the medial I in English is a marked option.[38] The German child also has both options, but her language favors the unmarked case.[39]

5.1. Wh-*movement and Copying*

We have, in effect, discerned a pattern in how the child assimilates primary linguistic data prior to a parametric decision. Does it occur elsewhere? We have uncovered a phenomenon that exhibits exactly the same kind of "copying" effect in *wh*-movement. We found that children would apparently answer the medial *wh*-expression in sentences of the form:

(59) a. how did she learn what to bake.
     b. when did he say how he hurt himself

The answer would be "cake" (59a), or "by falling out of a tree" (59b). It is notable that there was always a short-distance grammatical response available that made it unnecessary to build a long-distance chain. The child could always construe *how* in terms of *learn*. This pattern is described elsewhere (see deVilliers, Roeper, and Vainikka 1990, Roeper and deVilliers, 1991) and it has been replicated now several times. The phenomenon occurs primarily among 3—4 year olds and has been shown to occur with children in Dutch, German, French, and Spanish.[40] Children will also produce "copies", usually (but not always) with the same element:

"what did you say what it is" (three times in a row) and "I know something what you don't know what it is".[41]

One might object that the children merely hear the last part of the sentence in (59). We have performed several experiments which counter this possibility. First we found no tendency to treat a relative clause as a question:

(60)   How did the woman swim who knitted.

If children just listen to the last half of the sentence, then they should answer the question "who knitted". Second, if they ignored the "how", then they would have a yes/no question: did the woman swim who knitted. We virtually never received this response. Therefore the sentence-initial "how" is marking the utterance as a whole as a question. Second we tried the same sentence without *how* which produces a yes/no question:

(61)   Did she learn what to bake?

The children always answered "yes" or "no". This evidence then proves that it is precisely the chain in (62), the successive cyclic sequence, nothing else, which produces this effect:

(62)   wh . . . [$_{CP}$ wh . . . t]

Since the medial element is not a trace, then such sentences clearly deviate from the adult model of successive cyclicity.

These results are particularly noteworthy in light of several other facts. The children virtually never (eight examples from 1400 responses), allow the "when" to be interpreted in the lower clause if there is an intervening *how*. This is the basic claim of barriers-theory (Chomsky 1986) and it is clearly supported in our data.

The second notable fact is that, in a marginal way, the same behavior can be seen with adults when the *wh*-words are identical. The following sentence is seen to be ambiguous by many people:

(63)   now HOW did you say how you spun the canoe?

With emphasis particularly, it is natural to interpret this as a question about how the canoe was spun and not how one said it. The answer "loud" (how-said) is less likely than the answer "with two paddles" (how-spun). This must be a copying effect, otherwise it would be ruled out by the grammar under the barrier-theory.

Not surprisingly, there are grammatical systems in which this relation is a more prominent part of the system. In German, we find:

(64)   wie hat er gesagt, wie er malen wurde (how did he say how he
        would paint)

In addition, we find that there are cases of different *wh*-words marking each position:

(65)   was hat er gesagt wie er ein Kuchen backt
       [what did he say how he baked a cake]

These phenomena were briefly discussed by Van Riemsdijk (1983) and were then given an extensive description in the barriers framework by McDaniel (1989).[42]

What is the significance of these medial *wh*-expressions? The first observation is that they involve a minimal chain, where the notion chain refers to the number of traces. This straightforwardly fits a natural interpretation of the notion of "Least Effort" proposed by Chomsky.[43]

For adults, under current theory, there is a chain of the form: [wh . . . [$_{CP}$ tx[. . . V . . . t]]]. The first fact to note is that it is the non-lexical projection (tx) in CP which receives the copy, and never the object position of the verb. If the child had as its goal to articulate the origin of the *wh*-word, then the copy in the medial position would be very indirect. It is clearly the case that the child is articulating the chain itself.

The next question is really a prior question: what range of options does UG provide for long-distance questions? There appear to be several gross characteristics, although the theoretical literature is somewhat confounded by disagreements over data and theory. It is perhaps acquisition data of this kind which can make the parametric options distinct. In any case, roughly speaking, there are these options:

(66) a. Some languages have no LD-movement
     b. Some languages freely exhibit copies in medial contexts
     c. Some languages may allow LD-argument extraction without successive cyclicity.

The third case (61c) raises several suboptions: 1) direct long-movement in African languages discussed by Koopman and Sportiche (1985), or 2) arguments by Roeper et al. (1985) in behalf of the notion of a small pro object as a first representation of long-movement, 3) a predication relation between a *wh*-word, analyzed as a pronominal element, and a subsequent clause. This is advocated by Labelle (see below).

The default medial *wh*-word then establishes that pure long-distance movement is incorrect. Further evidence allows the medial *wh*-word to be replaced by a trace.[44] Before that point, unless the medial *wh*-word is present, it is not clear for the undecided child whether "long-movement" or successive cyclicity is involved. Therefore the medial *wh*-word disambiguates two parametric options. One important possibility should be noted at this point: it is possible that the default option is not evident in any adult language, but functions nevertheless for the child. This is an important possibility because it means that a synchronic analysis of adult languages alone cannot provide a full parametric map for children.

5.2. *French and* Wh-*movement*

Recent work by Labelle (1990) points out an interesting phenomenon found in French but not in English: the presence of clitic-copies for relative clause markers. She notes that French children, in large numbers, will add a clitic element (and occasionally a direct object):

(67)   "sur la balle qui l'attrape"
        [on the ball that he it catches]

She argues that these elements are present in French because there is a predicative relation between head nouns and relative clauses in French. This in turn is motivated by the presence of certain "predicative relative" expressions in French such as:

(68)   "C'est Marie qui arrive"
        [it is Marie who is arriving]

She argues that these forms which are common in adult and child French are not created by movement but form the basis of an overgeneralization that includes relative clauses.

Such expressions are quite acceptable in English as well and probably, though I have not searched it out, reasonably frequent in child language. In fact one example is in my corpus: "it's fish that we saw them" where the residual pronoun appears as she would predict. The connection to predicative structures is quite plausible, but it is less clear that it will serve as an explanation for another clear fact: French allows these "copies" in clitic position while English does not.

We think the explanation may lie elsewhere: in the fact that French in effect has two positions for direct objects: an object position and a clitic position.

(69) a. verb-obj: "sur la balle qui(l) lance la balle"
     b. clitic-verb: "sur la balle qui l'attrape"

If French and English both involve movement, why does the English child grasp the movement analysis immediately while the French child does not? Our proposal, once again, suggests that the unneeded clitic has a disambiguating function. Following Baker (1988) and others, we argue for a unique position for all theta-roles, though not necessarily a Maximal Projection. There are then two possible origins for the direct-object:

(70)   Clitic Parameter: The Projection Principle allows projection
        into:
        a) sister of V   = clitic position
        b) sister of VP = Object MP position

The role of the clitic is to select the clitic position as the origin in these
constructions. One might object that it is obvious that the true origin of
direct-objects in French is after the verb. However, it is quite plausible to
argue that children initially assume that all thematic roles are projected as
clitics and the full NP is treated as an extraposed copy. The NP-object
would then not be in an argument position. We will not provide a full
exploration of these options. We leave the matter with the observation that
there is also some surprising clitics in English acquisition data. Many
children will actually perform clitic doubling in English at a very early
stage even when there is no evidence for it (Brown 1973, Lebeaux 1988).

(71)   "close it the door"

The more general theoretical perspective is discussed by Jelinek (1984)
and Roeper and Keyser (1992) for language in general, and by Lebeaux
(1988) for acquisition.[45]

The question which then arises is this: why do they not occur in English? She suggests that it is because English actually has movement while
French does not. We would suggest two possible answers: 1) the crucial
factor is that there are two possible positions for the direct-object in
French and the child, and 2) the presence of clitics in the adult language,
like the presence of copies in the COMP in German, provides input for
marking the trace-chain. The clitic option, under this view, is abandoned
at an earlier point by the English child because there is no explicit evidence in the adult language to support it.

5.3. *Summary*

Although the analyses and data have both been intricate, the core of our
argument is simple and applies to all three environments. A copied
element disambiguates two possible chains — or locations for empty
categories — in the acquisition process. We suggest that this is a special
acquisition effect. It is a variant on Chomsky's notion of "saving a derivation" because the function of an phonetically explicit copy is to articulate a
chain which, given parametric alternatives, is otherwise opaque.

6. CONCLUSION

We have sought to put the acquisition evidence in a primary position in the evaluation of several linguistic proposals. Our goal here is to begin to develop a more refined theory of parameter-setting which can address the prior problem of how the primary linguistic data is analyzed and what UG must bring to bear to accomplish it. In that light we took, as evidence towards an acquisition theory, some very common examples of how children deviate from adult language in both production and comprehension.

6.1. *Goals: Biological Analogies*

How does acquisition evidence fit the larger goals of linguistics? The goal of linguistic theory is to provide an articulated formalism which can be interpreted biologically. What is relevant to a description of the biology of language? Heretofore the assumption has been that a synchronic description of adult competence should receive a unified biological explanation. This, however, does not exhaust the linguistic domain in biology. The acquisition of grammar must also receive a biological representation. In biology, a developmental role is often the full explanation of some feature of a phenotype. For instance, there is no non-developmental explanation in adult humans for the existence of belly-buttons. If I am not mistaken, there was an ancient school of thought that believed that the role of belly-buttons in adults was to define the center of symmetry in the body. This is the physical analogue to the presupposition that all features of grammar must have a synchronic explanation. This feature of adults has nothing more than a developmental explanation.

Some aspects of physical symmetry may receive a developmental explanation. If one examines other animals, we know that our ancestors once walked on four legs, not two. If one sees a child crawl, then the joint operation of arms and legs is crucial, even in the human being, though it is virtually not noticeable in adults. Crawling thus requires a symmetry which remains in the adult upright human, but which is utilized only by the child. Furthermore it triggers (by hypothesis) other capacities and may, it has been argued, even be related to the ability to read. A full understanding of our biology, then, makes important reference to the stage of crawling. Suppose there were no other animals who walked on four legs, and suppose we assumed that all early phases of human locomotion were different and therefore irrelevant. We would then miss an important piece of the biological puzzle. This feature has a marginal role for the adult, but a critical role for the child.

The growth of teeth provides a third biological analogy. In Roeper (1978) I discussed the idea that the instantaneous model might be wrong in subtle ways. For instance, looking at adult teeth one might imagine that they are grown versions of child teeth. In fact, however, an entire replacement process has occurred. Likewise acquisition might be non-instantaneous (therefore maturational); later grammars might replace earlier ones, rather than extend them.[46] Parameter-setting, which may reverse an early assumption, can be seen as a version of this model. This feature has a critical role for the adult, but its developmental origin is obscure.

The perspective we pursue in this essay is narrower: that UG is more than a map of parametric decisions. It entails analytic devices whose primary purpose lies in acquisition but which are still present in marginal ways in the adult language, as suggested by the analogy to crawling. Thus copying, like default case or adjunction[47] is important for the child in representing intermediate stages in the grammar from which further steps are possible.

NOTES

* This paper is an expansion of part of the paper (Roeper 1990) presented at Groningen 1989 at their 375th anniversary conference. It is substantially revised from the paper entitled "How The Least Effort Concept Applies to Head-Movement, Copying, and Wh-movement in acquisition" and it extends a paper entitled "How to Make a Parametric Choice" in Maxfield and Plunkett (eds.) 1991. Thanks for commentary to the members of the wh-project (Jill deVilliers, Tom Maxfield, Dana McDaniel, Jürgen Meisel, Ana Perez, Bill Philip, Berndadett Plunkett, Mari Takahashi, Jürgen Weissenborn) also to Noam Chomsky, Lisa Green Zvi Penner, Peggy Speas, Virginia Valian and audiences at Groningen, UMass, Leiden, and Budapest. Supported by NSF BNS-22574 and the Psycholinguistics Training Grant to UMass.

[1] We use the notation " *" to indicate unattested in acquisition corpora.

[2] Examples like (i) in contrast to (ii) indicate that *do* can be a proverb:

i)   what did you do   ⇒ play baseball
ii)   what did you want   ⇒ to play baseball

Play clearly substitutes for *do* rather than being a complement, this shows that *do*-insertion could apply to the main verb position as well as INFL.

[3] We would in fact question all supposed "performance effects" which show subtle grammatical sensitivity. One might in fact regard them in the same manner that one regards speech errors which also obey highly grammatical features.

[4] Ultimately when we understand the exceptional character of "construction-specific" rules, we may regard adult grammars as "inconsistent" in an important sense. The notion that lexical items can contain complex structural information allows us to have different lexical items whose structural characteristics are incompatible.

[5] See Lebeaux (1990) for discussion of the notion that a child's grammar may be stagewise inconsistent. See work by Crain et al. (1990) for evidence that different experiments elicit knowledge at different ages. We take any experimental results (as Lebeaux does) which show a deviation from adult response to be a clue to how a child attacks the acquisition problem: selection of a particular grammar. Evidence of stagewise inconsistency may nonetheless be important for applied linguistic perspectives.

[6] See Roeper (1981) for discussion of an input filter which has this function.

[7] See Borer (1984), Nishigauchi and Roeper (1987), Wexler and Manzini (1987), Clahsen (1989), Weissenborn and Verrips (1990) for a variety of proposals about lexical learning.

[8] This is, of course, a revived version of the principle of Structure Preservation proposed by Emonds (1976).

[9] See also Akmajian and Heny (1973), Menyuk (1973).
[10] See Roeper (1991) for extensive discussion of early stages of adjunction in English, Dutch, and German. Negation, modals, and adverbs all pass through a stage of pure adjunction where the syntactic category of the adjoined element remains underdetermined.
[11] See Borer (1984), Nishigauchi and Roeper (1987), Wexler and Manzini (1987), Clahsen (1989).
[12] This topic is considered in Pinker (1984) following joint work of Pinker and Lebeaux. The notion of "underspecification" is discussed there in terms of inflectional paradigms.
[13] Roeper corpus (1990).
[14] These facts altogether suggest that the notion of "save a derivation" does not capture all of what occurs. We shall not explore the theoretical implications further here, but simply suggest that *do*-insertion may also mark tense-scope for eventive verbs, creating a tense chain inside and outside the VP.
[15] The data assembled below is partly well-known data drawn from work by Davis and Mayer et al. (1976), partly from my own two children and searches of CHILDES. See Davis (1987) for a summary.
[16] Some examples of "do + have" are included, but this was not the focus of searches because of the complex nature of *have* in adult grammars.
[17] See Meisel (1985), Weissenborn (1987), DeHaan (1987), Pierce (1989), Clahsen (1989), Deprez and Pierce (1990), Tracy et al. (1990) for pertinent discussion of negation in French and German. Their evidence supports the notion that children have at least a finiteness node, a subpart of IP, from stages that are even earlier than those examined here.
[18] Again a closer analysis or more intensive diary search might be appropriate just at the moment when auxiliaries appear.
[19] In addition, a notion of "intervenor barriers" has been developed by Rizzi (1990) where he also argues that negation functions as an adverbial.
[20] It is apparently possible for a child to delete the Tense marker, which may reflect a stage before *do* is used to mark its position.
[21] See also Penner (1989) for discussion of *tun*-insertion in Swiss German. It is also commonly used in Dutch and German by both children and parents speaking to children "tue die Hände waschen" [do the hand (to) wash]. It therefore substitutes for V-2.
[22] The repetition of *do*-insertion in two clauses reveals that it cannot be a speech error in the classic sense, an on-line problem in retrieval of the main verb, since by the second clause the verb "ate" is certainly available.
[23] Tense-hopping may occur across sentence boundary: Adam "Was this is the boat I saw". See Phinney (1981) on cross-sentential neg-hopping.
[24] There are no examples so far of the form *'do he be sleeping* with progressive forms in the child language, though I am not sure that they are impossible. If so, then the non-raising phenomenon would be limited to main verb *be*. However, these progressive forms are possible in Black English.
[25] See M. Bowerman (personal communication), Roeper corpus (1990). See Meisel (1985), Slobin (1985) and references therein for the common view that aspect precedes tense. See Meisel (1990) for extensive discussion.
[26] The case of *have* is difficult to assess because of dialect differentiation. See Pollock (1989) for discussion which is also inconclusive.
[27] See Baker (1981) who claims "defective paradigms" do not allow generalization.
[28] If a lexical item is internally marked for tense, then, preferentially, insertion would occur after movement. The new unit $V + tense$ would then receive a single lexical item where both *verb* and *tense* are represented. This would lead to a natural constraint on parallel morphology:

    i)   Insert single words rather than compositional ones.
              [compositional = two morphologically separate items (verb, -ed)]

The converse markedness principle, favoring affix-lowering would be:

ii)   Only Phonetically real elements, not semantic features, are moved.

[29] Stromswold (1990) who has done a more exhaustive study and also found no instances of *"*does is* (personal communication). I have found one fixed idiom counter-example: "why do you're going outside", "why do you're giving juice", etc.

[30] The logic applies to *has* as well and we have not found examples with auxiliary *have* but some, predictably, exist for main verb *have*. Since *have* has extra complications it is not the primary example.

[31] Exceptions exist in the form of "don't he" in some contexts where there is an apparent lack of obligatory tense. However, there is dialect variation on exactly this point.

[32] The grammar poses other problems to the learner which could, in principle, complicate the picture. B. Plunkett (1989) points out that *be* is also associated with lowering in contexts like the following: *a boy is being bitten.* She (1991) also argues that the auxiliary *be* may be involved in simple sentences like *John is here,* for which the absence of *do*-insertion in acquisition provides immediate support.

[33] Note that participle formation has the superficial form of agreement as well: *John was pushed.* It could also provide input which would lead the child to believe that agreement between tensed *do* and tensed verb was natural. Thanks to P. Speas for pointing this out.

[34] This does not mean that the same communicative goal cannot be achieved by paraphrase.

[35] The same argument holds for sentences that involve several negatives. Properties of focus and emphasis, which we cannot characterize very well, are lost when we no longer say "No I am not a nothing boy".

[36] This can be read as a principled statement to the effect that a theory of acquisition cannot depend upon any form of evaluation Metric (see Chomsky 1965) since such metrics are in principle uncomputable.

[37] Special semantics may be preserved in the lexicon, as in "ain't" with a refusal reading.

[38] See Roeper (1990) for discussion of the possibility that English retains some SOV characteristics which could affect the acquisiton process.

[39] This might lead one to expect that there should be more copying in English than in German. This is my impression but it is difficult to document given uneven amounts of research in the two languages.

[40] See Maxfield and Plunkett (1991) for extensive discussion and details.

[41] From Roeper corpus. Numerous similar examples are reported by Crain and Thornton (1990).

[42] In various respects the child data departs from the behavior found in German. In German the wh-scope-marker is limited to *was*. It is natural to assume that in the learning process the child might exhibit a broader definition of scope marker.

[43] See also De Vincenzi (1989), and Pierce (1989) for further suggestions along these lines.

[44] See extensive discussion of these questions in the volume edited by Maxfield and Plunkett (1991). See also Crain and Thornton (1990) who provide a technical representation of these facts in terms of Spec-Head agreement. Although we believe that the medial *wh*-word represents the chain found in successive cyclicity, as we have argued, we do not believe that it has the properties of a trace. Therefore to label it a "trace-spellout" is misleading at best.

[45] See also Bresnan and Mchombo (1987) and the literature on many of the African languages.

[46] This perspective is discussed in terms of "discontinuity" in Pinker (1984) and in terms of maturation in Borer and Wexler (1987), Felix (1991), Clahsen (1991).

[47] See Lebeaux (1990) and Vainikka (1990).

REFERENCES

Akmajian, Adrian and Frank Heny: 1973, *In Introduction to Transformational Grammar*, MIT Press, Cambridge.

Baker, C. L.: 1981, 'Learnability and the English Auxiliary System', in C. L. Baker and J. McCarthy (eds.), *The Logical Problem of Language Acquisition*, MIT Press, Cambridge.

Berwick, Robert: 1985, *The Acquisition of Syntactic Knowledge*, MIT Press, Cambridge.

Borer, Hagit: 1984, *Parametric Syntax*, Foris, Dordrecht.

Borer, Hagit: forthcoming, *Parallel Morphology*, MIT Press, Cambridge.

Borer, Hagit and Ken Wexler: 1987, 'On the Maturation of Syntax', in T. Roeper and E. Williams (eds.), *Parameter Setting*, Reidel, Dordrecht.

Bresnan, Joan and Sam Mchombo: 1987, 'Topic, Pronoun, and Agreement in Chichewa', *Language* **61**.

Chomsky, Noam: 1965, *Aspects of the Theory of Syntax*, MIT Press, Cambridge.

Chomsky, Noam: 1986, *Knowledge of Language: Its Nature, Origin, and Use*, Praeger, New York.

Chomsky, Noam: 1986, *Barriers*, MIT Press, Cambridge.

Chomsky, Noam: 1988, 'Some Notes on Economy of Derivation and Representation', manuscript, MIT.

Clahsen, Harald: 1992, 'Constraints on Parameter-Setting', *Language Acquisition*.

Crain, Steven, Cecile McKee and Maria Emiliani: 1990, 'Visiting Relatives in Italy', in L. Frazier and J. deVilliers (eds.), *Language Processing and Language Acquisition*, Kluwer, Dordrecht.

Davis, Henry: 1987, *The Acquisition of the English Auxiliary System and Its Relation to Linguistic Theory*, dissertation, UBC.

De Haan, Ger: 1987, 'A Theory-bound Approach to the Acquisition of Verb Placement in Dutch', in G. DeHaan and W. Zonneveld (eds.), *Formal Parameters of Generative Grammar*, III-Yearbook 1987.

Deprez, Viviane and Amy Pierce: 1990, 'A Cross-linguistic Study of Negation in Early Syntactic Development', manuscript, Rutgers.

Disciullo, Anna Maria and Edwin Williams: 1987, *On the Definition of Word*, MIT Press, Cambridge.

Di Vincenzi, Marijke: 1991, *Parsing Sentences in Italian: Minimal Chain Principle*, Kluwer, Dordrecht.

Emonds, Joseph: 1976, *A Transformational Approach to English Syntax*, Academic Press.

Felix, Sascha: 1991, 'Universal Grammar and Maturation', in J. Weissenborn, H. Goodluck, and T. Roeper (eds.), *Theoretical Issues in Language Acquisition*, Erlbaum, Hillsdale, New York.

Flynn, Susan: 1987, *A Parameter-setting Model of L2 Acquisition*, Reidel, Dordrecht.

Frazier, Lyn and Jill deVilliers: 1990, *Language Processing and Language Acquisition*, Kluwer, Dordrecht.

Green, Lisa: 1990, 'Copular "be" in Black English', manuscript, UMass.

Hyams, Nina: 1987, *Language Acquisition and the Theory of Parameters*, Reidel, Dordrecht.

Jaeggli, Oswaldo and Nina Hyams: 1987, 'Morphological Uniformity', *NELS* **18**.

Koopman, Hilda and Dominique Sportiche: 1985, 'Theta-Theory and Extraction', GLOW talk, Brussels.

Labelle, Marie: 1990, 'Predication, Wh-movement, and the Development of Relative Clauses', *Language Acquisition* **1**(1).

Lebeaux, David: 1987, 'Comments on Hyams', in T. Roeper and E. Williams (eds.), *Parameter Setting*, Reidel, Dordrecht.

Lebeaux, David: 1988, *Language Acquisition and the Form of the Grammar*, Ph.D. dissertation, UMass, to appear, Reidel, Dordrecht.

Lebeaux, David: 1990, 'The Grammatical Nature of the Acquisition Process: Adjoin-a and the Formation of Relative Clauses', in L. Frazier and J. deVilliers (eds.) *Language Processing and Language Acquisition*, Kluwer, Dorerecht.

Maxfield, Thomas and Bernadette Plunkett: 1991, *Proceedings of the Umass Conference on the Acquisition of Wh-movement* [Available from UMass, GLSA, Linguistics Department].

Mayer, J. W., Ann Erreich and Victoria Valian: 1976, 'Transformations, basic operations and language acquisition', *Cognition* **6**, 1—13.

McDaniel, Dana: 1989, 'Partial and Multiple Wh-movement', *Natural Language and Linguistic Theory* **7**, 565—604.

Meisel, Jürgen: 1985, 'Les phases initiales du developpement de notions temporelles, aspectuelle et de modes d'action', *Lingua* **66**.

Meisel, Jürgen and Natasha Müller: 1990, 'On the Position of Finiteness in Early Child Grammar. Evidence from Simultaneous Acquisition of Two first Languages: French and German', manuscript, University of Hamburg.

Menyuk, Paula: 1973, *Sentences Children Use*, MIT Press, Cambridge.

Nishigauchi, Taisuke and Thomas Roeper: 1987, 'Deductive Parameters and the Growth of Empty Categories', in T. Roeper and E. Williams (eds.), *Parameter Setting*, Reidel, Dordrecht.

Penner, Zvi: 1989, 'The Acquisition of the Syntax of Bernese Swiss German: The Role of Functional Elements in Restructuring Early Grammar', manuscript, Bern.

Pesetsky, David: 1989, 'The Earliness Principle', paper presented at Glow, Utrecht.

Phinney, Marianne: 1981, *Syntactic Constraints and the Acquisition of Embedded Sentential Complements*, Ph.D. dissertation, University of Massachusetts, Amherst.

Pierce, Amy: 1989, *On the Emergence of Syntax: A Cross-linguistic Study*, Ph.D. dissertation, MIT.

Pinker, Steven: 1984, *Language Learnability and Language Development*, Harvard University Press, Cambridge.

Platzack, Christopher: 1990, 'A Grammar Without Functional Categories: A Syntactic Study of Early Swedish Child Language', *Working Papers in Scandanavian Syntax* **45**, 13—34.

Plunkett, Bernadett: 1989, 'Subject-Aux Inversion, That-t Effects and the Specifier Licensing Condition', *MIT Working Papers in Linguistics* **12**, 144—160.

Plunkett, Bernadett: 1991, 'Inversion and Early Questions', in T. Maxfield and B. Plunkett (1991).

Pollock, Jean: 1989, 'Verb Movement, Universal Grammar, and the Structure of IP', *Linguistic Inquiry* **20**(3).

van Riemsdijk, Henk: 1983, 'Correspondence Effects and the Empty Category Principle', in Y. Otsu et al. (eds.), *Studies in Generative Grammar and Language Acquisition*, International Christian University, Tokyo.

Rizzi, Luigi: 1990, *Relativized Minimality*, MIT Press, Cambridge.

Roeper, Thomas: 1978, 'Universal Grammar and the Acquisition of Gerunds', in H. Goodluck and L. Solan (eds.), *Papers in the Structure and Development of Child Language*, UMass, GLSA.

Roeper, Thomas: 1990, 'How the Least Effort Concept Applies to Head-Movement, Copying, and Cyclic Wh-movement', paper presented at Groningen, May 1989, UMass.

Roeper, Thomas: 1991, 'From the Initial State to V-2', in J. Meisel (ed.), *The Acquisition of Verb Placement, Functional Categories and V-2 Phenomena in German*, Reidel, Dordrecht.

Roeper, Thomas and S. Jay Keyser: 1992, 'Re: The Abstract Clitic Hypothesis', *Linguistic Inquiry* **23**(1).

104   THOMAS ROEPER

Roeper, Thomas and Edwin Williams: 1987, *Parameter Setting*, Reidel, Dorerecht.
Roeper, Thomas and Jill deVilliers: 1991, 'Ordered Parameters in the Acquisition of Wh-questions', in J. Weissenborn, H. Goodluck and T. Roeper (eds.), *Theoretical Issues in Language Acquisition: Papers from the Berlin Conference Hillsdale*, Erlbaum, Hillsdale, New York.
Roeper, Thomas and Jürgen Weissenborn: 1990, 'How to Make Parameters Work', in L. Frazier and J. deVilliers (eds.), *Language Processing and Language Acquisition*, Kluwer, Dordrecht.
Slobin, Dan: 1985, *The Cross-linguistic Study of Language Acquisition*, Erlbaum, Hillsdale, New York.
Stromswold, Karin: 1990, MIT dissertation, Psychology Department.
Thornton, Rosalind and Steve Crain: 1990, 'Levels of Representation in Child Grammar', paper presented at LAGB.
Tracy, Rosemary, 1991, *Sprachliche Strukturentwicklung*, Gunter Narr Verlag.
Tracy, Rosemary, A. Fritzenschaft, I. Gawlitzek-Maiwald and S. Winkler: 1990, 'Wege zur komplexen Syntax', manuscript, University of Tübingen.
Vainikka, Anne: 1990, 'The Status of Grammatical Default Systems: Comments on Lebeaux', in L. Frazier and J. deVilliers (eds.), *Language Processing and Language Acquisition*, Kluwer, Dorecht.
Weissenborn, Jürgen: 1988, 'The Acquisition of Clitic Object Pronouns and Word-Order in French', Max Planck Institute, Nijmegen.
Weissenborn, Jürgen and Maaike Verrips: 1990, 'Finite Verbs in the Acquisition of German, Dutch, and French', Max Planck Institute, Nijmegen.
Weverink, Meike: 1989, *The Subject in Relation to Inflection in Child Language*, Masters Dissertation, Utrecht.
Wexler, Ken and Rita Manzini: 1987, 'Parameters and Learnability in Binding Theory', in T. Roeper and E. Williams (eds.), *Parameter Setting*, Reidel, Dordrecht.

THOMAS ROEPER AND JILL DE VILLIERS

THE EMERGENCE OF BOUND VARIABLE STRUCTURES

1. INTRODUCTION

Even for adults, quantifiers such as "all", "some", "every" seem to involve a difficult mapping between logic and grammar. A sentence like "every boy ate every food" requires a little concentration before the meaning comes through. One might think that there is no natural mapping of such sophisticated aspects of cognition onto grammatical structure. Current linguistic theory, however, reveals that syntax puts sharp limits on how quantification works. The study of quantifiers might reveal how cognition connects to grammar and how they are intertwined in the process of acquisition. We will try to present the acquisition problem in a manner slightly abstracted from the technical details of linguistic theory.

The following two types of structures involve a quantifier which takes wide scope over a variable:

(1) a. Every boy sat on a chair.
    b. Dogs have a tail

The same interpretive option is available for *wh*-expressions which may be described in terms of bound-variables or in terms of pairwise connections at LF:

(2) a. Who is lifting his hat?
    b. Who brought what?

Each of these constructions involves a pairwise coupling, which can be called, descriptively, a bound variable reading (BV). In (2b) a pairwise answer is required. One cannot just say "people brought food" in reply, rather, one is obliged to say "Roger brought wine, Sally brought dip and Bill brought the quiche". In (1a, b and 2a) such a reading is optional: for instance in (1a) the boys could all be on one chair or each on his own. Each of the sentences in (1, 2) must have access to the BV notion in some form, and each is subject to different interpretive constraints.

Virtually no acquisition research has addressed the question of when these interpretations emerge in children's grammars. At the very least, the interpretations seem dependent upon a cognitive achievement, namely an ability to make pairings, or construct isomorphic correspondences. Piaget has argued that this notion of correspondence in the non-verbal realm is a crucial ingredient of intellectual growth in the preschool years. However, the linguistic bound variable reading consists in more than correspond-

Eric Reuland and Werner Abraham (eds), Knowledge and Language, Volume I, From Orwell's Problem to Plato's Problem: 105—139.

ences provided by the non-verbal context, because for the adult grammar, syntax places significant constraints on the bound variable reading.

An example of a syntactic constraint on quantifier scope is provided by relative clauses:

(3)   there is a horse for everyone
        (= each person has a different horse *or* one for everyone)

(4)   there is a horse that everyone is riding on
        (= one horse only)

One account of this difference is that, in order to produce the BV reading, "everyone" must move to a position (in Logical Form) outside of "a horse", which gives it "wide scope" over the NP "a horse". It is argued (May 1977, Chomsky 1986) that the principle of subjacency which prevents *wh*-extraction from relative clauses in the syntax also prevents quantifier extraction at LF.[2] This constraint would then apply to (4), but not to (3), thereby eliminating the BV reading for adults for (4). In other words, a structure of Logical Form must be generated by the child and syntactic constraints must be applied to that structure in order for the child to realize the distinction between (3) and (4).

A second illustration of a syntactic constraint on BV interpretation comes from the domain of "strong crossover":

(5) a.   whose hat is he lifting?
    b.   D-structure: he is lifting whose hat

In interpreting (5a) about a picture, a bound variable or paired reading is blocked: one cannot list the individuals who are lifting their hats, unlike the reading in (2a) above. It is possible to get accidental coreference if someone in the picture is lifting his own hat. But accidental coreference does not allow a set reading: a set of lifters and hat-owners that are connected. How can this block be explained? The *wh*-word functions like a name when it is c-commanded by a pronoun, preventing coreference, hence BV as well, as represented by (5b). But when do children know that the *wh*-word must be interpreted in its D-structure position?

A third illustration of a syntactic constraint is provided by the contrast (6a) and (6b) (May 1985):

(6) a.   who did everyone in our class marry?
    b.   who married everyone in our class?
    c.   someone married everyone in our class.

The question (6a) asks about pairwise couplings, while (6b) either refers to a minister, bigamist, or frequent divorcee. So the sentence (6a), where the *wh*-object is moved forward, allows a one-to-one, pairwise reading while such a reading is excluded for (6b), where the object remains in-situ.

In (6b) the "everyone" receives a "group" interpretation. Why should (6b) exclude the paired reading? It is not predictable from our ordinary understanding of events, that is, it is a syntactic block. It is notable that the restriction applies just to *wh*-words that have undergone movement: the example (6c) does not exclude the pairwise reading. How could a child learn to exclude the paired reading and allow only a group reading for (6b)?

In brief, May (1985) and Chomsky (1986) argue that the BV reading arises just when both variable elements are dominated by the same Maximal Projection. Universal Grammar requires that the *wh*-word automatically moves to the Spec of CP and thereby automatically acquires wide scope. The quantifier "someone" in (6c) optionally moves, at LF, therefore it is possible for the object "everyone" to move into a wide scope position and generate the distributed or BV reading. The logical question is: at what stage in development does this UG-requirement on *wh*-words become operative? We will argue that the availability of Spec of C may be crucial. We return to these analyses once we have laid out the empirical data.

Finally, consider a constraint on (7). Adults will readily get a bound variable reading when the plural NP is in the subject position, as in

(7)   Dogs have a tail.

In (7) we mean each dog has a separate tail, but we find it impossible when the plural NP is in object position:

(8)   A dog has tails.

In (8) the only meaning is that one dog has several tails. Thus the availability of the notion of pairwise correspondence is subject to subtle syntactic constraints.

There is considerable debate about the proper formulation of these constraints. Do they require a sophisticated and separate semantics, or a syntactic notion of Logical Form, or can they be captured within syntax itself? (See Chomsky 1986, Heim 1982, May 1977, 1986). These results must be addressed by any theory and therefore we pursue an exposition which, in part, abstracts away from particular formulations.

1.1. *Acquisition Issues*

There are three questions to address:

(9) a.   When do children show evidence of bound variable readings of linguistic stimuli?

b.   When do these bound variable readings become subject to syntactic constraints, and thus part of the child's grammar?

    c.   What empirical data and what grammatical decisions trigger the shift?

In the evidence assembled so far, it is clear that children realize the constraints on, for instance, "whose hat is he lifting?" and "there is a chair that every cat is on" at vastly different points, a difference of three to four years. Once the results have been presented, we return to the question of exactly how these constraints should be formulated.

    For just a glimpse of the magnitude of the acquisition problem, consider the points at which confusion could arise, given English data alone. A consideration of quantification cross-linguistically would further complicate the picture. For instance, one must have lexical knowledge of whether a quantifier is adverbial ("always") or nominal ("every"). In the following sentences it seems as if the two expressions are doing the same work:

    (10) a.   every person has a nose.
            b.   a person always has a nose.
            c.   some people have a nose
            d.   sometimes people have a nose

Suppose a child hearing (10a) mistakenly concludes that it was (10b), or hears (10c) and thinks it is (10d). Then "every" or "some" is an adverb which applies across a whole sentence and can appear anywhere in it. Why would she **not** come to this conclusion? Were she to come to this conclusion, she would then fail to see a distinction between: "Every official likes every talk" and "Officials always like always talking", or "Every cat likes every mouse" and "Cats always like mice". One could argue that there is a simple input which would work: a child could hear a sentence like "every boy likes every cereal" in a very clear context and determine from secure knowledge of the context that each quantifier must apply to each noun.[3] But such sentences, with clear contexts, are hardly frequent. A theory of acquisition must somehow guarantee that the child avoids confusion, and thus can lead to insight into the principles involved. It seems inevitable that the trigger is indirect: that the child learns to constrain the interpretation of "every" by locating it inside the NP determiner so that the scope restriction follows automatically. The child, in effect, must learn that articles and quantifiers are in complementary distribution in English (*"the every boy"). Then, of course, we must determine how the structure of the NP is acquired, given that the structure varies across languages.

    In what follows, a variety of experimental results with young children are discussed to attempt to determine the point of emergence of the linguistic notion of bound variable interpretations. After an initial overview, we return to discuss how different aspects of grammar are entailed by different structures.

2. EXPERIMENTAL STUDIES

In the first set of studies we explored children's answers to double *wh*-questions such as:

(11)   "Who ate which fruit?"

We contrasted that form with the subject *wh*-question (12a), and an echo-question (12b):

(12) a. "Who ate fruit?"
     b. "The family ate what?"

Recall that (11), for adults, requires a BV reading. And (12a) calls for either a group or variable answer of just the subject, although a BV reading is not ungrammatical. The echo-question (12b) calls for a literal repeat of the questioned word in the previous sentence.

We presented pictures to the child and a simple sentence such as: "The family ate fruit for dessert", then the question "who ate what?" or "who ate fruit?" or "the family ate what?"[4]

In Figure 1 it is clear that two individuals ate different things. We recorded what the child said and did, i.e., many of the responses were in the form of pointing, which we recorded as carefully as possible with a

Fig. 1.

videorecorder. We encourage the readers (especially those who have not done experiments) to think through each example carefully (saying the introductory sentences aloud) and, in effect, to perform the experiment on themselves.

There were various different logical responses that the subjects could make:

(13) a. give an **exhaustive paired interpretation** ("he ate this, and he ate that")
b. answer with **one pairing** (non-exhaustive) ("he ate that")
c. answer **generically**, e.g. ("The family ate fruit" or "fruit")
d. answer with an **exhaustive variable** interpretation of a **single** *wh*-question ("this guy and this guy")
e. answer with a **non-exhaustive**, singular interpretation of a **single** *wh*-question. ("this guy" or "an apple")

Adult-like behavior would entail giving the (13a) response to the instance: "Who ate what?" but NOT the answers (13c, d, e). This assumption was confirmed by experiments which we carried out in our classes. We performed pencil-and-paper versions of these experiments with at least 25 undergraduate students in each instance. The sentences were read aloud in the same fashion, but students would write the answer instead of saying it. We found that there was over 90% agreement on the adult answers.

Do children understand that a question word like "who" requires a variable response? Or is it treated like an empty name? (Equal to: "name a person that ate a fruit".) A response such as (13b) might indicate that the children failed to interpret the questions as variables, but it might also indicate a failure to master some pragmatic aspect of question/answer situations.[5] An absence of the notion of variable could also lead to answering the subject ("who ate fruit?") and object ("the family ate what?") questions with (13c)-type answers.

On the other hand, a BV answer to (11) or (12a, b) is not a grammatical violation in any instance.[6] We were initially seeking environments where we could elicit BV interpretations. Our results led us eventually, as we shall see, to explore syntactic environments where that reading is excluded.

Table 1 gives the incidence of the various types of answer for 17 children aged 4 to 6, and ten children aged 2.6 to 4 years. Notice that by four, the paired, exhaustive interpretation is well established specifically for the double *wh*-question (78.1%), and only six of the 27 failed to give any paired interpretations to this question type. Four of these children were among the five youngest in the group. Therefore it remains possible that at younger ages the BV reading is unavailable. On the other hand, the responses reveal that they seem to know the status of a *wh*-word as a variable. All but the two youngest children gave plural answers to

Table 1. 17 "old" children aged 4—6 years, 10 "young" children aged 2—3.11 years

Who ate which fruit?		Who ate fruit?	The family ate what?
a (bv) Old:	**78.1%**	**32.0%**	**30.3%**
Young:	**32.0%**	**57.1%**	**9.0%**
b (1—bv) Old:	1.5%	0.0%	3.0%
Young:	16.0%	7.0%	0.0%
c (gen) Old:	**9.0%**	**35.0%**	**33.3%**
Young:	0.0%	0.0%	0.0%
d (1/ex) Old:	9.4%	11.7%	20.5%
Young:	**41.0%**	**27.3%**	**54.5%**
e (1, nonex) Old:	1.5%	13.0%	3.0%
Young:	9.0%	7.0%	18.0%

(See 13a, b, c, d, e for interpretation of each type.)

questions at least some of the time, e.g. "the boy and the girl" or just "this one and this one".

In sum, by age four, the children have made a clear syntactic connection: the double-question structure must have a BV reading. However a surprising result appeared: the BV response occurred as one of the most frequent responses to a single *wh*-question (where adults would usually answer just the subject or object). It is clear that, when the BV reading is present, it is overgeneralized to contexts where it is, at least, pragmatically **unnecessary** for adults. For the group of children younger than four, the BV reading is linked equally to all three structures. What is the nature of such an overgeneralization? Does it imply that the bound variable readings are merely a cognitive strategy, or is there a syntactic representation in use by children that allows this extension?

The results warrant a close look. Generic responses (type c above) by age 4 were established for the single questions and very rare for the double *wh*-questions, which is precisely where we would argue they are forbidden by adult intuitions. In other words, children gave BV readings where we regard them as obligatory for adults. But there remains a puzzle: why do they extend the BV reading to cases where it is not obligatory (even if they are not ungrammatical)? The reader might want to say out loud the BV response to "who ate fruit?" to get an impression for the pragmatic overexplicitness of that response, in comparison to giving the straightforward answer provided earlier, namely "the family".[7]

Consider the other side of the coin: is there any domain where the BV reading is excluded? We sought a minimal pair for which the BV reading was obligatory in one case, and obligatorily blocked in the other. The pairs of sentences below (from May 1985) were used to see if children would select (a) for a group reading. While our other sentences called for the contrast between individual and a BV set, this called for the distinction between a group and a BV set:

(14) a. who pulled everyone?
     b. who did everyone pull?

As mentioned earlier in (6a, b), the paired reading is blocked for (14a), which has to mean: "who pulled the whole group?". In (14b) it is possible to get a distributed reading: "which person pulled each person?" We gave children four sets of pictures in which, for example, a series of people were pulling one another (see Figure 2) and asked two questions of type (14a) and two of type (14b) of each child.

Fig. 2.

Table 2.

Responses to Procedure 1[8] as a function of question type
(N = 16; ages 3;2 to 5;4)

	"group" answers	BV answers
Who pulled everyone?	25.5%	69.1%
Who did everyone pull?	11.2%	72.9%

Responses to Procedure 2[9] as a function of question type
(N = 19; ages 3;4—5;2)

	"group" answers	BV answers
Who pulled everyone?	77.2%	15.2%
Who did everyone pull?	73.5%	23.9%

We explored this contrast with several groups of children at the 3—4 year old range, varying the stimuli and the preamble in certain ways. The BV reading called for a pairwise articulation of what was happening (this one pulled this one, and this one pulled this one, etc.), while the group reading called for the children to point to the one character (he's pulling all the people). We found that the BV interpretation was overgeneralized again. The children were just as eager to take the BV reading for (14a) as for (14b) (See Table 2). This experiment therefore failed to find any syntactic limitation on the BV reading.

2.1. Wh- *and Indirect Questions*

Next we sought to see if the limitation would arise in contexts where indirect questions were asked. Indirect questions have the property that, being indirect, they do not seek answers, as in (15, 17):

(15)   Who did the father tell what to do?

(16)   Who did the father tell to do what?

and

(17)   Who did the father tell what to climb?

(18)   Who did the father tell to climb what?

Answer Types are:

a) BV: he told the girl to go on the swings and the boy to climb the slide.
b) single BV: he told the girl to go on the swings.
c) single, exhaustive wh: the girl and the boy.

In contrast (16, 18) are in situ questions which require a BV reading. Children were read a short story with accompanying pictures (see Figure 3), followed by one of the above questions. We imagined that in (15) the children should only answer the first *wh*-question, ignoring the question in complementizer position in the lower clause, as an adult would, because we do not answer indirect questions.[10] A subcategorized *wh*- in the CP is not bound as a variable to the fronted *wh*-word. In contrast, (16) requires a BV reading, as the *wh* is not in the CP. Table 3 shows that the 16 children in this study, aged 3;9 to 6;5, most frequently gave BV readings to **both** sentences, 15 to (15) and 14 to (16), with no distinction observed. Responses were slightly more distinct for the specific sentences (17) and (18), which avoided the generic proverb "do". In this case, 14 children gave BV responses to (17) but only 8 gave them to (18). Hence the phenomenon may be encouraged by certain aspects of the semantics of the sentence in question, but it is still present when these factors are minimized.

Fig. 3.

Table 3. Subjects: 16 children aged 3.9 to 6.5. Results: no. of children giving each answer.

	a	b	c	
15)	14	1	1	(tell what to do)
16)	12	2	2	(tell to do what)
17)	7	7	2	(tell what to climb)
18)	2	6	5*	(tell to climb what)

* 3 children answered "the slide", i.e., the medial question.

Further experimentation with adults, however, revealed that a surprising number of adults give a BV answer to (17). In fact, in Spanish this BV interpretation is an available part of the adult language.[11] The reader may note that if the "what" is stressed, the BV reading emerges more readily. Despite this residual effect in adult behavior, the fact remains that children were again projecting a BV reading where it is not required and not preferred. What does this imply? In order for adults not to answer the lower clause question, they must understand the question to be a subcategorization of a particular verb in English. Thus we have the following con-

trast: "He knew what he wanted" but not "*He supposed what he wanted".
We hypothesize that the children did not know that "ask" subcategorizes
for an indirect question. This is in fact confirmed by searches through the
naturalistic data (see deVilliers, Roeper and Vainikka 1990). Not knowing
the "what" to be a subcategorized indirect question, which calls for no
response, the children treat it as a real question, calling for an answer. In
that regard, it is equivalent to an in situ case like "He supposed that he
wanted what"? which is perfectly grammatical without a special subcategorization. Under the broad assumption that lexical learning is slow, it is
predictable that the children turn to the BV reading. We return to the
question of how the subcategorization arises at a later point.

We argue that the BV interpretation in these cases is made possible by
the child's grammars, in which the subcategorization has not yet been
established for "ask" — Q. There is a further dimension of difference
between medial and in situ questions which helps establish the precise
syntactic limitations governing the children's interpretations. Sentences
with in situ wh-words cannot have the wh-word function as a barrier.
Consider a second important difference between (19) and (20).

(19)   How did the girl choose t what to wear *t?

(20)   How did the girl choose t to wear what t?

As mentioned, the first difference is the topic of discussion: wh-in situ
(20) calls for a distributed BV response. The second difference depends
upon the theory of barriers: if an adjunct "how" is moved, it cannot pass
through a CP with another wh-word present (19). (In technical terms,
following Lasnik and Saito (1984) proper government is required for the
intermediate trace which does not occur if there is a branching node
present).[12] This effect does not hold for (20) with wh-in situ. Therefore we
can interpret (20) as "how-wear" and not just "how-choose".

In fact we found very clear evidence that children do not allow "how"
to move over "what" in CP: among a group of 16 4–6 year olds, 36%
allowed long-distance interpretations for (20), while only 5% allowed
them for (19).[13] If children are sensitive to this barrier effect, then it
follows that they are aware that "what" is in the Complementizer position.
But the results on "ask" questions indicate that they are apparently
unaware that "ask" and now "choose" are also lexically subcategorized to
allow an indirect question. If not an indirect question, then "what" must be
interpreted as a real question. One way to make it a real question is to
give it a pairwise multiple wh-interpretation together with "how".[14]

In sum, we have located a syntactic barrier effect, but failed so far to
find a constraint on the BV interpretation. The importance of this result is
that it shows that a very precise syntactic awareness is at hand: children
are apparently not free to use ordinary inference in interpreting wh-

questions in complex environments. Their interpretations are subject to tight syntactic constraints. This suggests (but does not prove) that if the BV reading is available, then a specific grammatical analysis must allow it. Our goal is to provide such an account rather than to assume that the children's interpretations fall outside of the grammar.

However, if we find no context in which BV analysis is disallowed, then a grammatical explanation is weak. Roeper et al. (1984) found just this kind of evidence: children reject the BV interpretation for single-clause strong-crossover sentences. Children between 3—9 years were given a picture that had two possibilities: two Sesame St characters, each lifting their own hats, and one person lifting Big Bird's hat. The experiment was replicated a number of times with different age groups. They were then given sentences of the form:

(21) a. Nl: who is lifting his hat?   (36.9% = BV)
     b. Cl: whose hat is he lifting? (3.6% = BV)

      N = non-crossover, C = crossover
      1 = one clause, 2 = two clause

(See Figure 4.)

Fig. 4.

Neither sentence elicited large numbers of BV readings, particularly from the youngest children, while two clause sentences elicited around 30%:

(22) a. N2: who thinks he is lifting his hat   (38.1% = BV)
    b. C2: who does he think is lifting his hat (29.8% = BV)

The results above are from a group of 21 children 5—7 years. In addition with a group of 22 children 3—5 years, we found comparable results from a set of 528 sentences, 63 of which received BV interpretation (see Table 4):

Table 4.

	N	C
1 clause	7	4
2 clause	26	26

Again it is clear that the single clause cases strongly resist BV interpretation, and that two clause BV interpretations were in the minority.[15] The results suggest that children are able to reconstruct a trace in the single clause sentences which, as illustrated above in (5), rules out a coreferential reading.[16] If we combine this result with our extensive evidence of freely available BV readings, it suggests children in this age range are sensitive to at least one of the adult restrictions.

Let us now summarize what we have observed so far about BV in *wh*-contexts. We have examined three contexts where BV is disallowed for adults:

(23) a. Object quantifier: who pulled everyone?
    b. Subcategorization: who did you tell what to do?
    c. Cross-over: whose hat is he lifting?

In the third, we find the constraint obeyed, suggesting that children are able to recognize a D-structure empty category and make the appropriate interpretation. In the second case, we have argued that subcategorization is missing. In the first case, note that a quantifier is present. Before interpreting the quantifier case, we turn to a detailed examination of quantification structures where there is more evidence of BV overgeneralization.

3. QUANTIFIERS

3.1. *Quantifiers and Subjacency*

Do children respect the possibly clearer linguistic constraints on non-

Fig. 5.

question quantifiers? We created pictures (see Figure 5) that depicted several possible interpretations of sentences such as:

(24)   Every child sat on a horse

(25)   There is a horse that every child sat on

and we asked the child to choose the right picture to go with our sentence, from one depicting each child on a different horse, one depicting all the children on one horse, and one showing three children on their own horses and one without a horse. 21 children aged 3.7 to 7 years demonstrated that they allowed both interpretations readily for (24), and almost equally readily for (25). That is, 12 of the 21 children gave us a BV reading for (25), even asking on occasion, "Do you mean one, or a lot of horses?" Once again, syntactic structure had no impact: the BV response was overgeneralized to include subjacency environments where it should be excluded.

We replicated that study using a slightly different methodology in which we ask the child a truth-judgement question about a single picture, for sentences such as:

(26)   Is there a chair that every cat is on?

(27)   Is every cat on a chair?

15 children aged 4.3 to 5.7 participated, and demonstrated the same intuitions as the previous subjects: 11 of them accepted 100% of the pictures for (26) in which each cat was on a different chair. In fact, three children rejected the reading of (27) for a picture in which all the cats were on one chair, saying:

"No, there's only **one** chair"

This is clear evidence that the notion of BV is overgeneralized, and often strongly preferred over the narrow scope reading of (27). Our results in this domain replicate similar findings by Lee (1986) who did comparable experiments in both English and Chinese.

In sum, we have failed to find syntactic limitations on quantifier interpretations, and this is reminiscent of the overgeneralization of bound variable readings for *wh*-questions.

3.2. *Plurals*

In a pilot study (carried out by Anne Vainikka) with 15 children aged 3;7 to 6;0, children were asked a variety of questions of the following sorts (no child received two questions with the same content):

(28) a. Do dogs have tails?
     b. Does a dog have a tail?
     c. Do dogs have a tail?
     d. Does a dog have tails?

Animals and animal parts were varied (*Does a cat have noses?*). To our consternation, the children showed no differentiation among the four types: the answer was almost always "yes". We also explored the issue in more informal conversations. In the pre-school period, the answers are uniformly positive. In this domain too, although we have yet to explore it systematically, the BV reading was overgeneralized.

3.3. *Quantifiers and Indefinites*

These results are reminiscent of a famous, but never explained, result obtained by Donaldson and Lloyd (1974). They gave children a picture of four garages, three of which were filled with cars, with one empty. They asked the children:

(29)   "Are all the cars in the garages?"

Surprisingly the children pointed to the empty garage and said "no, this is empty". The "all" appears to apply to both cars and garages and the goal seems to be, once again, an isomorphic (or BV-like) connection between cars and garages (see Philip and Takahashi (1991) for discussion).[17] We have dubbed this phenomenon "quantifier-spreading":

(30)   Q-Spreading: A quantifier attached to one NP applies to all NP's in a clause.

This result, in turn, finds support in work by Roeper and Matthei (1974) with the quantifiers "some" and "all", who suggested that quantifiers initially have an adverbial character. Children between the ages of four and six years were asked to interpret the sentence:

(31)   Some of the circles are black

They were given a set of pictures to choose from (see Figure 6), and they frequently chose a picture where some of the circles were partially black IV. In other words, they interpreted one "some" as if it applied to both NP's, "some of the circles are some black", just as "all" does, and just as "every" apparently did for the children above. This is a crucial, but predictable, consequence of the hypothesis that children detach quantifiers from the nouns they appear with. What kind of input could support such an analysis?

3.3.1. *Quantifier-Float*

The kinds of "all" structures that children receive is worth a moment's reflection. Unlike "some", "all" undergoes what is called "quantifier-float",[18] moving like an adverb away from the noun it modifies:

(32) a. all boys like chocolate
     b. boys all like chocolate
     c. all boys are now here
     d. boys are now all here

(33) a. some boys are now here
     b. *boys are now some here

It appears then that it is a simple accident that *some* does not float, a possibility that warrants a careful cross-linguistic study.

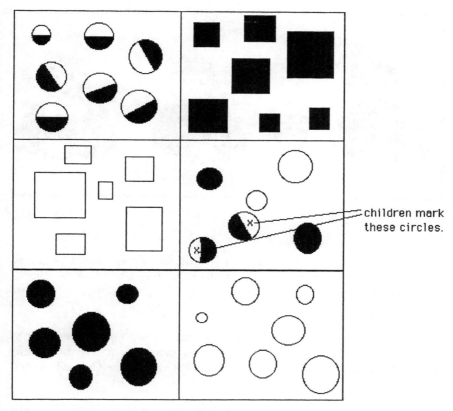

children mark
these circles.

Show me a box where some of the circles are black.

Fig. 6.

3.4. *Quantifiers and Blocked Spreading*

William Phillip and Sabina Aurelio replicated the Donaldson and Lloyd result with, e.g. Figure 7 and the question:

(34)   Does every boy have a milkshake?

74% of the time children between 2—5 years responded "Not this one" while pointing at the extra milkshake. How general is this phenomenon? In particular, is this a linguistic or a cognitive phenomenon? By analogy with the arguments above for BV readings in the adult language, we would expect to find syntactic constraints operating on "Q-spreading" if it were a linguistic phenomenon. The central question is this: what boundaries exist for spreading? There are three sentential contexts in which Philip and Takahashi have tested to see if children would still permit spreading.

Fig. 7.

A) If c-command ("every" over "a") is required for spreading in "every body is drinking a milkshake", then no spreading should occur "backwards" (in "a" over "every" sentences):

(35)   A cat is on every chair

If spreading goes backwards (35 = every cat is on a chair), then it does not obey c-command. Instead the quantifier can move forward to dominate all NP's, much like the movement of a PP in "into the garage I pushed the car".

B) If Subjacency is a barrier, then spreading should not occur from an NP outside a relative clause to one inside (36 = every whale is lifting every boat):

(36)   Every whale that is lifting a boat smiled

or from inside a relative clause to outside (37 = every waiter is carrying a glass):

(37)   A waiter who is carrying every glass is falling down.

If spreading occurs also into and out of relative clauses, then some feature of subjacency is not present.

C) If no second NP is present then syntactic spreading should not be possible.

(38)   Every dog is sleeping.

Such sentences were presented to children together with a picture involving dogs sleeping on beds, with an extra bed in the picture. If the beds go unmentioned in the intransitive sentence (38) then the children should not point to the extra bed and say: "not this one".

3.4.1. *Backwards Spreading*

In several studies, we have varied the position of the quantifiers to see if there is any effect of linear ordering on the spreading phenomenon. Quantifier spreading was just as likely in these contexts, suggesting that c-command is not a necessary constraint on its appearance.

3.4.2. *Spreading and Relative Clauses*

Pictures like Figure 8 were used with relative clause sentences.

Children again showed a strong inclination toward spreading over the relative clause boundary, just as we found with the "there"-insertion constructions in section 2.3 above. When asked "Is every whale that lifted a boat smiling", they answered "No, not that one", pointing to a boat.

The children were significantly less likely to spread in relative clause environments, showing that they did detect a difference in the structures,

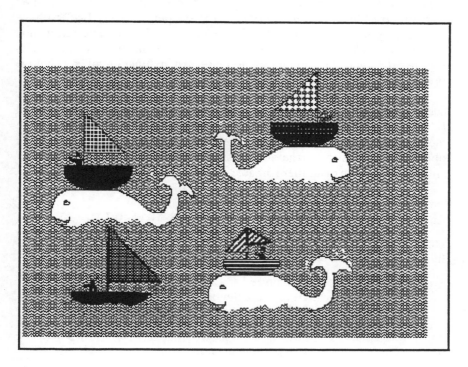

Fig. 8.

but the spreading phenomenon was still strong enough to represent a
marked violation of subjacency as a barrier.[19]

3.4.3. *Spreading and Intransitives*

If spreading were to occur to an unmentioned, but pictured object, then
one could make an argument that the phenomenon was linguistically
unconstrained. If however the spreading occurs only in environments
where both NP's are mentioned, then it is clearly linked to what the child
understands the language to allow. We refer the reader to Takahashi
(1991) where English and Japanese experiments are discussed in which
children exhibit the spreading phenomenon for sentences like (39a) but
not (39b):

> (39) a. A cat was climbing every ladder
> b. Every dog was sleeping
> c. Every boy was driving. A truck was broken.

The children would say "no" to (39a) approximately 50% of the time and
point to a cat climbing a tree and say "not this one". They would answer
"yes" to (39b) even though one bed was occupied by a cat. Had they
understood (39b) to mean "every dog was sleeping in every bed" to have
the meaning [every [bed and dogs]], then we would have expected a "no"
answer. We take this as evidence that the phenomenon is, minimally,
sensitive to syntactic limitations and not just a function of a cognitive
preference for isomorphism lying outside the grammar.

In fact, Philip and Takahashi (1991) have uncovered a particularly
subtle contrast between two kinds of intransitives (39b) and (39c). In
(39c) there is the possibility of an implicit object, unlike (39b). They in
fact find that children will overgeneralize with respect to the implicit
object during a certain stage of acquisition: "every boy is driving (a truck)"
and once again the quantifier spreads.

4. QUANTIFIER-SPREADING AS ADVERBIAL

How shall we analyze the phenomenon of quantifier-spreading? Our basic
hypothesis is this:

> (40) a. Quantifiers are analyzed as adverbs
> b. Adverbs can be given sentential scope
> c. Therefore all NP's within a clause are modified by the adverb.

Two other studies point in the same direction. The well-known phenomenon of Neg-hopping is a comparable phenomenon:

> (41)   He doesn't think John ran = he thinks John did not run

In the current analysis of barrier-theory, Rizzi (1990) specifically argues that Neg functions as an adverb. Evidence from Phinney (1981) showed that children are more liberal than adults in allowing Neg-hopping. It is restricted to a few verbs for adults, but not for children. In an experiment she showed that children consider the sentence "the bears saw the children not eat honey" to be the equivalent of: "the bears did not see the children eat honey."[20] This is, once again, just as if the child allowed a negative-adverb to take scope over the entire sentence.[21]

Consider now the experiments with plurals in which we found that children consistently answered sentences of the form "Does a dog have noses" with "yes". The answer fits an analysis where plural and negation both receive a kind of "concord", the plural spreads from one NP to another just as negation spreads (suggested to us by B. Schein). We taken this to be a description of a process whereby an adverbial operator is attached outside the highest node:

(42)

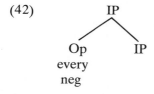

In current theory, it is often suggested that there is a NegP position at the top of the IP. We would suggest that it needs to be defined more broadly as an AdvP node where all adverbs that modify any element below the CP are defined.

The concept that plurality can function as an operator, and therefore allow movement, is built into the notion that plural agreement is possible. In effect, instead of saying that a plural marker, i.e., the AGR node, moves between an NP and a verb, this suggests that the Operator can be attached to the sentence as a whole initially. Growth consists in making a specific adjunction to a verbal head. It is notable that current theory has argued for a separate node for each of these elements: AgrP, NegP, and QP under DP (though not for plurality). (See Pollock 1989, Speas 1990, and references therein.) We will not explore the theoretical implications of this observation at this time, but focus instead on the acquisition perspective.

Consider now the original scene: "every car is in a garage" means that a car is in every garage. Note, however, that we cannot simply copy the "every" into two Spec positions. The sentence does not have the impossible meaning (43):

(43)   every car is in every garage.

It is some equivalent of the form "every car and garage has 'infulness' ". It is not easy to characterize the translation of meaning here, but it is clear that "every" must apply to a joint vision of cars and garages in order to avoid the absurd reading in (43). Consider again the facts presented by Roeper and Matthei (1974) who proposed that quantifiers can be analyzed as adverbs. In fact the reading of "every" is very close to the adverbial reading linked to the word "always" and "some" to the meaning of "somewhere", each of which has sentential scope. Imagine the meanings for: "somewhere the circles are black".

4.1. *Naturalistic Evidence*

There is evidence from naturalistic data to support this hypothesis as well. A number of children have been recorded or reported as saying:

(44)   Maria: "Only I want this one" (meaning: I want only this one)
                "Even I want you to drive me to school"
                "I only can walk"
                "I just only have a hood" (= only I have a hood)
                "I can even not believe how hot my back is"

         Adam: "Only go dere"
                "Only hit Adam"

         Sarah: "Only people do this"
                "Only take one"
                "Only it doesn't have, like that."
                "Only I see 'R'."
                "Only if I put this. . . ."
                "Only start from the corner."
                "Nope, just only hot dogs, too."

The expressions "just" and "even" will often show up misplaced in children's sentences, sounding comprehensible, but slightly odd to adults.[22]

4.2. *Free Adjunction Hypothesis*

Several kinds of evidence then point to the hypothesis that quantifiers may be analyzed as adverbs, and that adverbial quantifiers that are linked to NP's will be misplaced as sentential modifiers. The adverbial interpretation also fits the notion of "free adjunction" suggested by Lebeaux (1988) as a default property of grammars:

Default: Adjoin new material to the highest node possible.

That is, children can freely attach adjuncts to higher nodes to represent new input. The concept of a default means that when new analyses arise, they are automatically preferred. Therefore, a revision in NP structure may create the possibility for a lower attachment. This will automatically eliminate the higher attachment or make it a marked case. In this sense, **free adjunction** is a default operation. This theory of defaults fits the suggestion by Chomsky (1988) that the language-particular properties of grammar remain distinct from universal properties. Once again, if a language particular analysis arises, the UG default analysis is automatically abandoned.

4.3. *Spec Variation*

Some fundamental questions still remain unanswered: Why should quantifiers be analyzed as adverbs and how does the child find her way out of this false generalization? Before we proceed, let us cast our net wider, and see if there is relevant cross-linguistic evidence.

The quantifier-adverb hypothesis has recently received support from other work in linguistics. Work by K. Hale (personal communication) and E. Bach (personal communication) indicate that in widely diverse languages, there is always an adverbial quantifier available, but not always a quantifier as a determiner.[23] Under the "default" approach one predicts that the adverbial reading may continue to be available in marginal constructions in English. One finds, in fact, such an implication in the sentence:

(45)   John saw another hitchhiker down the street, so he went to a different corner.

The implication of the word "another" is that John is also a hitchhiker [another [John and hitchhiker]]. In some languages, according to Hale, all quantifiers are construed with this kind of sentential scope, just as we described for "every" above. Thus diverse evidence supports the view coming from acquisition that quantifier = adverb could be the unmarked hypothesis.

What must the child acquire in order to use quantifiers in English appropriately linked to an NP? Note that the behavior of quantifiers in NP's in English is not uniform. Consider just this variation:

(46) a. all the boys
     b. *every the boys/*some the boys
     c. some of the boys
     d. *every of the boys
     e. the boys all
     f. *the boys every

It is clear that each quantifier has special lexical characteristics which must be learned: "all" can appear with a full NP in pre-NP or post-NP position, "some" can appear with a PP complement, "every" cannot co-occur with a determiner. It is possible that some of these differences are linked to semantic differences. Nevertheless, there are a number of distinctions that the child must correctly identify.

There are two possible rationales for a child's initial misanalysis: 1) an unmarked analysis is taken, and 2) the adult analysis is unavailable. These factors conspire in the acquisition process in a way that is not fully understood. In other words, the reason that a default analysis is chosen is not simply because it is unmarked, therefore preferred. The reason is that a logically prior decision has not been made. Once the language-particular analysis is secure, the default analysis disappears. In this instance, we argue that it is the absence of a full NP structure which pushes the child toward an adverbial analysis.[24]

The following hypothesis is advanced:

(47)   Quantifiers are adverbs until the Spec of NP is fixed.

This hypothesis can be interpreted in at least two ways which need refinement in our future work. The first possibility is that the Spec of NP is absent at first, and only when it is triggered, can quantifiers be appropriately accommodated within the NP. (If we assume that there is a Determiner Phrase which dominates the NP, then it will be the Spec of DP which must be fixed, in order to allow quantifiers, which was suggested to us initially by William Philip.) In support of that claim would be the argument that Japanese, for instance, does not have any Spec of NP. If language variation exists, any particular Spec node must be triggered.

A second possibility is that the Spec of NP already exists, but that the lexical variation described above means that each quantifier has to be separately justified as belonging to some node in the NP, and before that, each quantifier is analyzed as an adverb. Obviously some combination is also possible: first no Spec, then separate justification quantifier by quantifier. We hope to examine the acquisition of each quantifier in order to approach this issue in a more refined way. Interestingly, these precise alternatives also present themselves in considering the *wh*-question analysis too (see below).

The adverb analysis we have presented does not differentiate the syntactic and semantic components. We have argued simply that the child makes an adverbial analysis of quantifiers because their syntax is incomplete.[25] It can be argued that properties of quantifiers come not from their syntactic categorial features, but from the fact that they can raise to sentential level (via QR) and bind NPs — just like adverbs.

(48)   If John sees a milkshake, he drinks it.

This has been analyzed by Heim as involving a quantificational adverb at the semantic level:

(49)   ALWAYS$_{-j}$ [John sees [a milkshake]$_{-j}$] he drinks it$_{-j}$

This semantic analysis does not depend on the presence of a real syntactic adverb. In other words, the entire analysis could be projected at the semantic level. We believe, however, that the child arrives at the correct, restricted analysis by virtue of an interaction between possible syntactic representations and their semantic representation. Shifts in the syntactic representation then entail shifts in the semantic representation. The discussions by Takahashi and Philip provide a careful description of a changing "restrictive" clause in a semantic representation which could lead to such an analysis. Their analyses, like this one, assume that it is changes in the syntactic representation which restructure the available interpretations.

Another way to view the phenomenon is to assume that the adverbial interpretation is limited to a comprehension representation. Takahashi (1991) documents that while children are able to use nouns like "everybody" there is virtually no use of "every" in forms like "every boy". Could the child have access to the meaning of individual words and then directly to an LF structure without ever forming a syntactic structure? We could expect this response to remain available as a default even among older children who have begun to use "every" within the Determiner Phrase and therefore have the ability to project the needed syntax. It would remain in the child's grammar for a period of time as a fairly rare structure, just as in the adult language the adverbial interpretation of "another" remains as a marginal possibility.

5. SPECIFIER AS MP TRIGGER:
CONNECTING ADVERBS AND *WH*-QUESTIONS

So far we have provided an empirical discussion and a theoretical claim about how quantification emerges. Is there a way to unite this discussion with the observations about *wh*-interpretations? We turn now to a broader acquisition theory in proposing the following hypothesis:

(50)   Hypothesis:
   1. Heads do not automatically project Maximal Projection nodes.
   2. Corollary: the Spec node must be specifically triggered for each MP.
   3. Parametric variation: some Spec nodes are optional.

In particular, the Spec node of NP and CP are our focus here. It is possible that similar arguments can be made about Spec of IP and VP.

Our hypothesis is a specific version of the general claim that functional categories are delayed in emergence in child grammar, proposed by Lebeaux (1988), Guilfoyle and Noonan (1988), Radford (1989), Platzack (1990). These claims, in turn, fit the claims in linguistic theory, e.g. Fukui and Speas (1986), that there is a wide proliferation of XP categories, with considerable variation. We provide here just a summary of the highlights of these arguments. The term 'delayed in emergence' is chosen carefully. The delay is often taken to be maturational, but we do not construe it in this fashion, although maturational factors could in principle also be involved. Given the diversity of languages, certain kinds of evidence will be differentially available at different times. Therefore it is not surprising if the CP node is available immediately in German, but not for a long time in English.[26]

Our proposal is simply that functional categories require specific triggers. Those triggers are more or less opaque depending upon the language. A language where all quantifiers are uniformly to the left of the NP will be easier to acquire than a language where a quantifier, like "all", can appear on both sides. Suppose the child projects a general phrase structure rule of the form: $[Q - N]_{NP}$. The quantifier appears before the noun. However, she hears sentences like: "the boys all came". This sentence conflicts with the phase-structure rule. So what does she do? A) abandon the rule as wrong, B) add a new phrase-structure possibility, C) add a new transformation, D) avoid the phrase-structure rule until both that rule and a transformation are generated. Under any choice, the acquisition pattern will have to be more complex than in a language where all quantifiers behave identically and do not move.

5.1. *Lexical Aspects of Spec*

Roeper (1988) initially proposed that the Maximal Projections NP, VP, CP, and PP were each triggered by the emergence of a Spec.[27] One can get a feeling for the general claim by considering PP's. Many languages allow *wh*-preposing in PP's. In German this a productive operation. Most of the starred cases below are acceptable. English allows *wh*-preposing in PP's in only a limited fashion, except for a few residual cases, usually with "where":

(51)   *howunder, *whoin, *whenfor, *whyby

(52)   whereby, wherein, whereto,
        ?wherefrom, ?wherefore,
      *?whereunder,
       *wherewith (but "wherewithal"),
       *whereabout (but whereabouts),
       *wherenear, *wheretoward, *whereamong.

One can argue that there is no Spec in PP's, but rather the acceptable forms have been lexicalized.[28] The child, despite hearing a few cases in (49) must not make the false generalization of Spec in PP.[29] (See Roeper and Weissenborn (1990) for discussion of the problem of avoiding false generalizations because of lexical exceptions.[30])

5.2. *Absence of Spec = No Maximal Projections*

Now let us consider two hypotheses: the Spec node is absent in CP and in NP. The absence of a Spec node in the CP would mean that the CP was not a Maximal Projection. There are many consequences to this claim. We list a few here, which are discussed elsewhere in greater depth:[31]

(53) a. An absence of inversion of auxiliaries in children's *wh*-questions
    b. Copying of the initial and medial *wh*-word in children's grammars[32]
    c. An absence of subcategorization of indirect questions.

The absence of auxiliary inversion in acquisition is one of the most well-known phenomena that has been studied, e.g. children say

(54) "what you are doing?".

One feature of non-inversion has come in for less discussion:[33] it persists until six or seven with certain *wh*-words, generally "why" ("why he can't eat"), while it disappears with others. This, by itself, indicates that *wh*-words might be separately justified as belonging in Spec of CP, and that only when they are in Spec of CP is there the opportunity for the auxiliary to move into the head of CP.[34]

What feature of grammar allows optional movement until this point? Here we find the parallel options to the quantifier case above. One possibility is that the Spec of CP is absent at first,[35] just as we argued that the Spec of NP may be initially missing. Second, the Spec of CP may be present but the *wh*-words each require justification as belonging in that position as opposed to some adjunct position, and until they are so justified, they remain as adjuncts to IP. A possible trigger for the reanalysis that we have suggested is the appearance of the *wh*-word appropriately subcategorized in the medial CP, which de Villiers (1991) has reported as being strikingly coincident with the emergence of inversion in the matrix clauses for each *wh*-word. A third possibility, as before, is that these are stages: first no Spec, then Spec justified for each *wh*-word in turn. At the very least, it seems that the Spec node in CP remains optional for some period in childhood. As a consequence, the usual claim that the significant fact is when children begin using inversion (with some falling back to earlier grammars) is altered: the significant moment is when inversion becomes obligatory.

5.3. *Quantifiers and There-insertion*

Can we apply this notion of Spec as an optional node to clarify any of the findings above? Recall that adults, but not children, will block a BV reading for sentences like "there is a chair that every cat is sitting on". This restriction has been assimilated to the subjacency constraint on extraction. It is noteworthy, however, that Otsu (1981) demonstrates the presence of subjacency at the level of S-structure, namely for *wh*-extraction, as early as 3 years:

(55)   What is the woman painting a bird that flew with?[36]

That is, children will not misconstrue (55) as referring to the long wings the bird flew with. Yet even seven year olds are making mistakes with the quantifier case of extraction from relative clauses. The developmental difference, then, is enormous. It is, moreover, not the case that children do not have long-distance movement at this stage. Our evidence clearly indicates that successive-cyclic movement must be present (see de Villiers et al. 1990), complete with barriers to movement. If another *wh*-word occupies the medial CP, then that serves as a barrier to successive cyclic movement of the initial *wh*-word:

(56)   How did the boy ask t when to jump *t?

Therefore a CP must be available. Why should children have this extra degree of freedom in quantifier extraction?

The optional-Spec concept leads to the prediction that children will allow quantifiers to move over CP barriers. If quantifiers are being analyzed as adverbs, they do not undergo Successive Cyclic Movement at LF, and hence do not cycle through the CP. Nevertheless, they are subject to subjacency restrictions, namely, they cannot cross Maximal Projections. But on our analysis, if the Spec is not present, then the CP is not a Maximal Projection, just a C, hence, not a barrier. Therefore the adverb-movement is not blocked by a non-maximal C.

It is important to note that the above argument uses both the quantifier-as-adverb hypothesis and the optional-Spec hypothesis in order to account for all of the child's behavior. By the time the child is six or seven, they are generally able to use quantifiers within NP's. Therefore the quantifier is no longer an adverb. However, in the formation of an LF representation, the quantifier moves together with its N to determine scope. The absence of a Maximal Projection CP node would then allow the true quantifier to have wide scope over the NP which is directly dominating it:

(57) a.   there is [a chair [$_C$ that every cat is sitting on t]]

[every cat [a chair]]

unlike the case in adult grammar with [$_{CP}$:

(57) b.   there is [a chair [$_{CP}$ that every cat is sitting on t]]

5.4. *Copying and Spec*

A striking finding about young children's interpretations of sentences containing two *wh*-words:

(58)   How did the boy ask what to bake?

is that the children answer the medial *wh*-word almost as frequently as the initial *wh*-word (de Villiers, Roeper and Vainikka 1990). The facts can be accommodated under the theory that the medial *wh*-word is not at first interpreted as belonging in the Spec of the medial CP, and only when it is so interpreted is this co-indexation between the initial and medial *wh*-word disallowed, and long distance (successive cyclic) movement is then possible (subject to barriers). The phenomenon is also consistent with the facts on subcategorization, discussed next.

5.5. *Subcategorization and Spec of CP*

There are two dimensions to subcategorization. 1) The child must decide which verbs take complements, and in particular, indirect questions. 2) The child must decide which *wh*-words are questions and which are adverbs, and which are both. Note that some *wh*-words have a referential function as well as a question function. A sentence with an adverbial conjunction "when" does not cause inversion because it is not a question: "when I came home, I had a sandwich".

If the Spec of CP is the ultimate position where indirect questions must be, then it is predictable that inversion, subordination, and the triggering of Spec of CP will all co-occur. This then fits the framework we have outlined.[37]

5.6. Wh- *and Wide Scope*

We turn now to the question of why sentences like "who saw everyone" initially receive a misanalysis, allowing wide scope for "everyone".[38] Movement to Spec of CP at LF guarantees wide-scope for "who" and narrow scope "everyone" which in turn produces the group reading (a):

(59) a. $[_{CP} [_{spec} who_i [_C +wh] [_{IP} [_{spec} t_i [_{VP} [_V saw everyone]]]]]]$

b. $[_C [_{IP} [_{spec} who_i [_{VP} [_V saw everyone]]]]]$

If however, there is initially no Spec of CP (59b), then this would enforce the non-movement of *wh-* at S-structure and lead to the prediction that either *wh-* or "everyone" could receive wide scope at LF, just as we find with "someone saw everyone". All of these diverse arguments, summarized briefly here, point to the possibility that children would lack the Spec of CP.

We do not, of course, regard the current theory of CP as immutable. There are many cross-linguistic issues to be addressed before we can be confident of how complementation systems work. Our argument provides a particular slant on a general problem in linguistic theory. In brief, children allow a broader interpretation of quantifiers and a narrower interpretation of *wh-*extraction (copying) at the same time. Any future theory of constraints on complementation must address itself to these facts as well as the cross-linguistic ones.

6. PARAMETERS AND PRIMARY LINGUISTIC DATA

Acquisition theory has two distinct tasks: 1) to explain the instantiation of UG, and 2) to trace the map of parametric choices. The latter task has been, recently, built into UG itself under the assumption that UG will describe a set of choice points addressed by the child. Therefore it is often asserted that Universal Grammar is equivalent to an acquisition device or equivalent to the initial state of the grammar.

The parametric problem has held the focus of attention during the last decade: how does the child select the one grammar, among all those defined by UG, that fits the language around him? The effort to make a parametric map has not been obviously successful. In each instance where a decision point is defined, one can point to acquisition data or language variation which could confuse the child and create precisely the indeterminacy which the parameter was intended to eliminate. This suggests that special principles of acquisition may be needed which define certain data as primary.

In Roeper and de Villiers (1991) we discussed the fact that certain decisions must be linked to a unique trigger.[39] For instance, the child must regard the sentence a) "what did he do?" as signalling a *wh-*movement language although he hears and uses routine forms like b) "you know what?" and hundreds of echo questions like c) "he did what?" Such facts (b, c) should trigger English as part of the *wh-*in situ language family; or the combination (a, b, c) should leave the child in a state of utter indeterminacy. We have some evidence which suggests that initially children do permit in situ interpretations of echo-questions, which shows that rejection of the in situ option is not straightforward.[40] One conclusion is that the parametric map by itself will not exhaustively define the principles needed for acquisition, but rather, acquisition principles that are not visible within synchronic grammars by themselves, will be needed to guarantee that available data does not mislead children.[41]

6.1. *Conclusion*

A fundamental linguistic distinction — distributed (BV) versus non-distributed readings — in a variety of linguistic contexts, has been the focus of this study. The notion itself seems intuitively sophisticated from both a cognitive and linguistic point of view. Yet our studies have shown that it was cognitively available at a young age and initially overgeneralized. The operative assumption here is that children's behavior in this domain must be compatible with and licensed by their grammars, which therefore requires a linguistic rather than an extra-linguistic explanation.

The evidence from acquisition and cross-linguistic work argues, thus far, in behalf of one primary claim: children treat quantifiers adverbially. This claim, in turn, has been cast within a broader acquisition theory: the Spec nodes of certain categories are delayed in emergence. The delay arises because the combination of syntactic and semantic data the child encounters lends itself to misanalyses which, historically, has been regarded as the fundamental acquisition problem. The solution lies in identifying unique triggers: *wh*-movement to the clausal periphery (Spec of CP) may be such a trigger, affecting ultimately, not only *wh*-movement but quantification.

Much remains to be done. We need a detailed map of the emergence of quantification. It will undoubtedly lead to more insights into the Adverb-Hypothesis and recast our view of the Spec-hypothesis.

NOTES

[1] Anne Vainikka, Sabina Aurelio, William Philip, and Mari Takahashi, have been crucially involved in carrying out many of these experiments; several of their more extensive empirical and theoretical discussions will appear elsewhere. In addition, our whole *wh*-acquisition group has contributed advice at all levels. They include Bernadette Plunkett, Dana MacDaniels, Tom Maxfield, Meike Weverink, Fei Xu, Ana Perez-Leroux, Anne Vainikka, Jürgen Weissenborn, and Juan Uriagereka. Anne Vainikka carried out the experiments on plurals. Jill Van Antwerp carried out the experiment on "who pulled everyone", and drew wonderful pictures. We have also benefitted from comments at various presentations at BU, Groningen, Leiden, and UMass. Comments by S. Crain, J. Frampton, B. Schein, B. Partee, P. Portner and M. Speas among others have been helpful.

[2] There is a good deal of controversy over this claim (See Lasnik and Saito (forthcoming) and references therein). Although there has been counter-evidence, recent work (Nishigauchi 1987) points again at subjacency effects. We take, at the minimum, the subjacency formulation as a description of restrictions on the interpretation of quantifiers in relative clauses.

[3] See Roeper (1981) for discussion of this approach. Also see Hornstein and Lightfoot (1981) for discussion of "exotic triggers".

[4] Each child received 4 BV questions, 2 subject and 2 object questions, but with no two questions about the same story.

[5] For instance, there are some, not so common, adult environments where it is acceptable to give less than exhaustive replies. If we ask "where can I sit" one does not have to name

every chair. But, on the other hand, if we ask "who was in the car", we would err in failing to mention someone. We believe that the exhaustive reading for questions is clearly the grammatical requirement, with a few pragmatic exceptions.

[6] In point of fact, we have found that full sentence responses are much more common among children than constituent responses (although more work is involved). Whereas adults prefer to answer the question "what did you eat" with "cookies", children generally respond with "I ate cookies". We are preparing a more extensive study of this question.

[7] One might object that pragmatically new information is sought rather than the repeat of old information. This is just not true in the life of a small child. Large parts of the dialogue between parents and children are of the form: "this is a washing machine" followed by a test question "what is it", where the child says "a washing machine". The child is showing not only that he "knows" the obvious answer but that he can pronounce the words, which may be a more significant and rewarding challenge.

[8] The first procedure consisted in giving the full story: e.g. "This little boy was out in the country one day when he got stuck in the mud. His sister tried to pull him out but he was really stuck. Then the Dad came and tried to pull the sister but it was no use. Then a horse came along and pulled the Daddy and look! Out came the boy!"

[9] In the second procedure, we tried to balance the preamble to de-emphasize the pairings: we told the same story, and ended it with: "So the horse pulled this long line of people and this long line of people pulled the boy". Clearly we were too successful!

[10] The reader might note that there are contexts in which we answer the indirect question: Do you know what time it is? However, when asked a question like "Can you always see what you want on TV?", there is no real answer to the wh-word. If children answered the wh-word, the question would be noticeably misunderstood.

[11] Pointed out to us by Ana Perez-Leroux.

[12] We present a more complete discussion in de Villiers and Roeper (in preparation).

[13] See also de Villiers, Roeper and Vainikka (1990).

[14] Production data indicates that indirect questions are acquired verb by verb and wh-word by wh-word, see below.

[15] We refer the reader to Roeper et al. (1984) for extensive discussion. The adult responses were found with a group of 8—10 year olds.

[16] The fact that the strong crossover sentence (d) does not rule out the BV reading then leads to an interpretation of both cases in terms of small pro. The small pro in the single clause cases would rule out BV readings under Principle B, while BV would be allowed in for the two clause cases because Principle B no longer applies.

[17] This phenomenon is the focus of work by Bill Philip, Sabina Aurelio, and Mari Takahashi, who provide a discussion of experiments and a more extensive theoretical interpretation. We present here our initial work on the topic with some references to their forthcoming work. (Phillip and Aurelio 1991, Phillip 1991, Phillip and Takahashi 1991, Takahashi 1991.)

[18] Sportiche (1986) and others have argued that it is not the quantifier that moves, but the N moves away from the quantifier. We use the terminology of "quantifier-float" although these results are equally compatible with the other view.

[19] Phillip and Aurelio also constructed examples containing a discourse relation where quantifiers were involved with indefinites, and not pronouns. This work is still undergoing refinement and, and though there is support for our view of the constraints on BV interpretation, we will only allude to their results here. They gave children a picture with chickens and eggs in baskets. They then asked the children to say if the following statement was true:

(i)   Every chicken stood up. An egg hatched.

In other examples, the direction of the quantifier was reversed:

(ii)   A dog got on a bed. Every cat jumped.

Only one child out of 12 exhibited spreading in these environments.

[20] These results were obtained in a similar manner: an array of pictures depicting all logical possibilities was presented.

[21] This fits the notion that verbs are initially underdefined so that they can function as bridge verbs more easily, which we have discussed elsewhere (de Villiers, Roeper and Vainikka 1990, Roeper and de Villiers 1991). Verbs also fail to subcategorize properly, as we mentioned above: children do not initially see that *ask* takes an indirect question, and many other verbs as well. Therefore the broad phenomenon of how the meaning of verbs emerges, their potential subcategorization, and the potential for long-distance movement over clause boundaries all develop together.

[22] Nina Hyams (personal communication) has also noticed phenomena of this kind.

[23] See recent work in *Papers on Quantification* (Bach, Kratzer, and Partee 1989) for some of the interpretive complexities of adverbs.

[24] The difference between comprehension and production is important here. In effect, the child may primarily use the adverbial analysis in comprehension. Comprehension is forced not only in our experiments, but in many contexts where universality is the essence of a conversation. "you ate all the dessert" has meaning in a situation where "you ate dessert" has no consequences.

[25] We are indebted to John Frampton for discussion.

[26] This point is perhaps due a moment's reflection. Consider the example of inflections. They are equally evident on a phonetic level in different languages. A child can hear the -s in *he runs* as easily as a German child hears the -t in *er läuft*. In English inflections are notoriously confused, while in languages with consistent inflections, they are not. Jaeggli and Hyams (1987) argue that there is a morphological parameter which, in the unmarked case, requires no inflections or a full paradigm of inflections. Languages with incomplete paradigms are hard to learn under their parametric definition because the child must assemble a fair amount of evidence to determine that they belong to neither class. What does this mean? It means that the data is clear in a minimal sense, but the parametric setting is more obscure in one language than another. The delay has to do with the obscurity of the parameter.

[27] We have not extended the argument to consider the range of new MP's currently proposed for the IP complex. In general, the more hypotheses there are that in some languages a particular notion is treated as an affix, while in others it is an MP, the more natural it becomes to argue that MP structure requires specific triggers. In effect, there is a three way variation: affix, head, MP. The potential status of affixes, of course, complicates this picture and takes us far beyond the scope of this paper.

[28] This argument, like most, has further complexities. One must account for intensifiers in PP's as well: "far under", "all the way out", "more near" etc. Suffice it to say that the Spec of PP, should it exist, would require a definition that allowed a non-productive *wh*-preposing.

[29] See Roeper and Weissenborn (1990) for discussion of how a child deals with contradictory data. Also Clahsen (1990).

[30] The structure and acquisition of IP is a topic of great controversy. Several authors have argued that the child's initial subjects are generated within VP (Pierce 1989, Clahsen 1990). One can, in fact, argue that initial stages of inversion in yes/no questions are merely apparent inversion, because the child leaves the subject in the VP. This stage would be supported if the child did not initially have a Spec node for the IP. Clahsen (1990) advances just such an argument claiming that the child developmentally moves from X^0 to XP. We argue that the shift has a very precise character: emergence of Spec in each of the MP's.

[31] See de Villiers and Roeper (1990a, b).

[32] An absence of the movement of complex *wh*-phrases "which hat" in early stages would be a natural corollary of this prediction under the hypothesis that only Spec of CP allowed MP's. This is a theoretically controversial domain, however, since Lasnik and Saito operate without a Spec of CP and other languages with putatively no Spec must then allow MP's to occur in Comp.

[33] But see Kuzcaj and Brannick (1979).

[34] See de Villiers (1991).

[35] It may be argued that the whole CP is absent at first, but by the time this is occurring, there is overwhelming evidence for CP in children's grammar: all manner of embeddings and auxiliary inversion in yes/no questions.

[36] We have recently explored adjunct *wh*-extraction from relative clauses and found even more striking obedience to subjacency with 3 year olds. See de Villiers and Roeper (1991).

[37] Lebeaux (1988) has advanced the hypothesis that *wh*-questions could at first be generated in situ in the COMP position.

[38] This is akin to Chomsky's initial argument (extended from George 1980) that children might, at S-structure, fail to move a subject *wh*-expression. Such movement is obligatory at LF.

[39] See also Roeper and Weissenborn (1990).

[40] This work is still in preparation. See Takahashi (1991) and Maxfield (1991) for experimental results showing a variety of contexts in which children are sensitive to echo-questions.

[41] This is, in effect, a translation of the original formulation of the acquisition problem which led Chomsky to formulate an evaluation metric and an instantaneous model.

BIBLIOGRAPHY

Bach, Emmon, Angelika Kratzer and Barbara Partee (eds.): 1990, *Papers on Quantification*, UMass.

Chomsky, Noam: 1986, *Barriers*, MIT Press, Cambridge.

Chomsky, Noam: 1988, 'Some Notes on Economy of Derivation and Representation', *MIT Working Papers* **10**, 43—74.

Clahsen, Harald: 1990, 'Constraints on Parameter-setting', manuscript, University of Düsseldorf.

de Villiers, Jill, Thomas, Roeper and Anne Vainikka: 1990, 'The Acquisition of Long Distance Rules', in Lyn Frazier and Jill de Villiers (eds.) 1990.

de Villiers, Jill and Thomas Roeper: 1991, 'Introduction', in T. Maxfield and B. Plunkett (eds.) 1991.

de Villiers, Jill: 1991, 'Why Questions', in T. Maxfield and B. Plunkett (eds.) 1991.

Donaldson, M. and P. Lloyd: 1974, 'Sentences and Situations: Children's Judgment of Match and Mismatch', in F. Bresson, *Problemes Actuel en Psycholinguistique*, CNRS.

Frazier, Lyn and Jill de Villiers: 1990, *Language Processing and Acquisition*, Kluwer, Dordrecht.

Fukui, Naoki and Margaret Speas: 1985, 'Specifier and Projection', *MIT Working Papers in Linguistics* **8**, 128—172.

George, Leland: 1980, *Analogical Generalization in Natural Language Syntax*, Ph.D. dissertation, MIT.

Guilfoyle, Eithne and Maire Noonan: 1989, 'Functional Categories and Language Acquisition', unpublished manuscript, McGill University.

Heim, Irene: 1982, *The Semantics of Definite and Indefinite Nounphrases*, UMass Ph.D. dissertation, GLSA, Amherst.

Hornstein, Norbert and David Lightfoot: 1981, 'Introduction', *Explanation in Linguistics*, Longman, London.

Jaeggli, Osvaldo and Nina Hyams: 1986, 'Morphological Uniformity and the Setting of the Null Subject Parameter', *NELS* **18**, GLSA, Amherst.

Kuczaj, Stanley and N. Brannick: 1979, 'Children's Use of the Wh-Question Modal Auxiliary Placement Rule', *Journal of Experimental Child Psychology* **28**, 43—67.

Lasnik, Howard and Mamoru Saito: forthcoming, *Move Alpha*, MIT Press, Cambridge.

Lebeaux, David: 1988, *Language Acquisition and the Form of the Grammar*, UMass Ph.D. dissertation, GLSA, Amherst.

Maxfield, Thomas: 1991, 'Children Answer Echo Questions How?', in Thomas Maxfield and Bernadette Plunkett (eds.) 1991.

Maxfield, Thomas and Bernadette Plunkett (eds.):1991, *Papers in the Acquisition of Wh*, University of Massachusetts Occasional Paper in Linguistics. GLSA, Amherst, MA.

May, Robert: 1977, *The Grammar of Quantification*, MIT Press, Cambridge.

May, Robert: 1985, *Logical Form*, MIT Press, Cambridge.

Nishigauchi, Taisuke: 1987, 'Subjacency and Japanese', manuscript, UMass.

Otsu, Yokio: 1981, *Universal Grammar and Syntactic Development of Children*, Ph.D. dissertation, MIT.

Philip, William: 1991, 'Event Quantification in the Acquisition of Universal Quantifiers', manuscript, UMass.

Philip, William and Sabina Aurelio: 1991, 'Quantifier Spreading in the Acquisition of *Every*', in Thomas Maxfield and Bernadette Plunkett (eds.) 1991.

Philip, William and Mari Takahashi: 1991, 'Quantifier Spreading in the Acquisition of *Every*', in Thomas Maxfield and Bernadette Plunkett (eds.) 1991.

Phinney, Marianne: 1981, 'Children's Interpretations of Negation in Complex Sentences', in Susan Tavakolian (ed.), *Language Acquisition and Linguistic Theory*, MIT press, Cambridge.

Pierce, Amy: 1989, *On the Emergence of Syntax: A Cross-linguistic Study*, Ph.D. dissertation, MIT.

Platzack, Christer: 1990, 'A Grammar Without Functional Categories: A Study of Early Swedish Child Language', manuscript, University of Lund.

Pollock, Jean-Yves: 1989, 'Verb Movement, Universal Grammar, and the Structure of IP', *Linguistic Inquiry* **20**, 365—424.

Radford, Andrew: 1989, *Syntactic Theory and the Acquisiton of English Syntax*, Blackwell, Oxford.

Roeper, Thomas: 1981, 'The Role of Universals in the Acquisition of Gerunds', in E. Wanner and Leila Gleitman (eds.), *Language Acquisition: The State of the Art*, Cambridge University Press, New York, 267—287.

Roeper, Thomas: 1988, '*Wh*-Movement and Spec Acquisition', paper presented at Boston University Conference on Language Development.

Roeper, Thomas and Jill de Villiers: 1991, 'Ordered Decisions and the Acquisition of *Wh*-movement', in Helen Goodluck, Thomas Roeper, and Jürgen Weissenborn (eds.), *Theoretical Issues in Language Acquisition*, Erlbaum, Hillsdale, New York.

Roeper, Thomas and Jürgen Weissenborn: 1990, 'How to Make Parameters Work', in Lyn Frazier and Jill de Villiers (eds.), *Language Processing and Acquisition*, Kluwer, Dordrecht.

Roeper, Thomas, Mats Rooth, Lourdes Mallis and Satoshi Akiyama: 1985, 'On the Problem of Empty Categories in Language Acquisition', manuscript, UMass.

Speas, Margaret: 1990, *Phrase Structure Syntax*, Kluwer, Dordrecht.

Sportiche, Dominique: 1988, 'A Theory of Floating Quantifiers and Its Corrollaries for Constituent Structure', *Linguistic Inquiry* **19**, 425—449.

Takahashi, Mari Bernadette: 1991, 'Children's Interpretation of Sentences Containing *Every*', in Thomas Maxfield and B. Plunkett (eds.) 1991.

Vainikka, Anne: 1990, 'Comments on Lebeaux', in Lyn Frazier and Jill de Villiers (eds.), *Language Processing and Acquisition*, Kluwer, Dordrecht, 83—103.

M. RITA MANZINI

CATEGORIES IN THE PARAMETERS PERSPECTIVE: NULL SUBJECTS AND V-TO-I*

The primary aim of this work is to provide an account for the null subject and V-to-I parameters and for their interaction. The data considered are the standard ones in Rizzi (1982) for null subjects, and in Pollock (1989) for V-to-I (in finite sentences). Our discussion is articulated in three sections. After an introduction to parameters and phrasal dependencies in section 1, our version of the V-to-I parameter is presented in section 2, our version of the null subject parameter, together with our conclusions, is presented in section 3.

1. BACKGROUND

Two classical models of the null subject parameter are presented in Chomsky (1981) and Chomsky (1982), following Jaeggli (1982) and Rizzi (1982) respectively. According to the first of these two models the possibility of having null subjects depends on the possibility of moving I to V in the syntax, as in (1). Suppose however that, as proposed in subsequent literature, a constraint holds to the effect that parameters correspond to properties in the lexicon. The version of the null subject parameter in (1) can then be ruled out, in that it violates this principle:

(1)   *Null Subject Parameter (1)*
        I can/cannot move to V in the syntax

The idea that parameters correspond to lexical properties, first explicitly suggested in Borer (1984), is formalized in Wexler and Manzini (1987) as the Lexical Parameterization Hypothesis, according to which values of parameters are associated with lexical items. Thus, as argued in Manzini and Wexler (1987), different values of the locality parameter for referential dependencies are associated not with different languages, but with different lexical items within the same language. Assuming that the intuition is correct, we can attempt to reformalize it as the Lexical Parameterization Principle in (2). This says that values of parameters are features, i.e., grammatical classes, that have as their members either single lexical items or else other grammatical classes, i.e., categories:

(2)   *Lexical Parameterization Principle*
        Values of parameters are grammatical classes with lexical items
        or other grammatical classes as their members

Consider now the second classical version of the null subject parameter.

141

Eric Reuland and Werner Abraham (eds), Knowledge and Language, Volume I, From Orwell's Problem to Plato's Problem: 141—155.

This does not suffer from any problem with respect to (2), since the possibility of null subjects depends on the possibility of I being associated with the feature + pronominal, as in (3). (3) also has the right properties in other respects. In particular, the feature it makes use of is an independently known feature, which is furthermore known to vary independently of the category I:

(3)   *Null Subject Parameter (2)*
        I can/cannot be [+pronominal]

Next, Pollock (1989) argues that languages differ as to whether they allow movement of a main verb and/or an auxiliary to I. In Chomsky's (1989) terms, movement of V to I is allowed or not allowed according to whether I is strong or weak respectively. In other words the parameter in (4) arises:

(4)   *V-to-I Parameter*
        I is strong/weak

In what follows we will first review what we take to be the crucial properties of phrasal expletives and then return to our account of the V-to-I parameter. We adopt the theory of indicization in Manzini (to appear) under which all lexical categories and only lexical categories are associated with what we call a categorial index. (Categorial) indices are indices of lexicality, as in (5); in the case of lexical D/N heads they are meant to subsume standard referential indices. The definition of chain, or the definition of sequence that underlies it according to Manzini (to appear), is formulated in terms of c-command and compatible indexing, rather than of coindexing, as in (6):

(5) α has a (categorial) index iff α is lexical

(6) $(\alpha_1, \ldots, \alpha_n)$ is a sequence iff for all i
        α_i c-commands and has an index compatible with α_{i+1}

Within our theory, then, we postulate the existence of categories that have lexical content and categories that are non-lexical. Formally, lexical categories are represented by (categorial) indices, and nonlexical ones by their absence. If an expletive is lexical we must associate it with an index; at the same time we propose that its dependent status is represented by this index being a variable one. Thus a typical expletive sentence is associated with the indexed structure in (7). We can assume that a variable index has exactly the same properties as the absence of an index from the point of view of sequence formation; in other words, being compatible with all indices, it can enter a sequence with any of them. This means that *there* and *a woman* can form a sequence in (7) as is generally assumed:

(7)   There$_x$ arrived a woman$_i$

It is worth noticing that in terms of the theory that we have developed an anaphor or pronoun must also be associated with a variable index, exactly like an expletive. Hence the morphological identity of expletives and pronouns is not coincidental. The variable index of anaphors and pronouns allows them to participate in sequence formation, exactly as the variable index of expletives. The difference between an expletive and a referential dependency seems to be produced simply by the position of the variable index with respect to the non-variable one. Referential dependencies however have no direct place in our discussion of heads.

2. THE V-TO-I PARAMETER

The standard data concerning V-to-I movement are presented in Pollock (1989). The adverb *completely/complètement/completamente* cannot follow (and must precede) a finite V in English, but it must follow (and cannot precede) it in French and Italian, as in (8)—(9). Let us assume that the base generated position of the adverb is adjoined to V'. Following Pollock (1989) the data in (8)—(9) are explained if finite V must remain in place in English and move to I in both French and Italian. As a consequence, the adverb will show up to the left of V in English, but to its right in French and Italian:

(8)   En.   a. *John lost completely his mind
        Fr.   b. Jean perdit complètement la tete
        It.   c. Gianni perdette completamente la testa

(9)   En.   a. John completely lost his mind
        Fr.   b. *Jean complètement perdit la tete
        It.   c. *Gianni completamente perdette la testa

The question is why the head movement patterns individuated in Pollock (1989) and only those are instantiated. What we want to suggest is that the crucial property of English in this respect is that its I is lexical; while the crucial property of French or Italian is that their I's are non-lexical. Thus the relevant parameter is as in (10). The idea is that if I is indexless in French and in Italian, V can move to I exactly like a phrase can move to an empty position; while V cannot move to I in English exactly like a phrase cannot move to an already filled position. If in English the index that I is associated with is a variable index, on the other hand, under our definition this of course allows for the formation of (I, V) expletive sequences:

(10)   *(V-to-I) Parameter*
        α is/is not lexical

Consider French or Italian. The theory actually accounts for the fact that

V can move to I. In particular, movement of V to I yields the structure in
(11). In (11), the index of V percolates up to I, which is indexless, so that I
binds the trace of V; the trace of V is then licenced by I. As for the fact
that movement must take place, we can adopt the account in Chomsky
(1989). According to this, if no movement takes place the resulting
configuration is excluded for morphological reasons, relating to the affixal
nature of I. Similarly following Chomsky's (1989) account, the reason why
I movement to V does not represent a possible alternative to V to I in
French or Italian is the Least Effort Principle; we can adopt this either in
its original version or in Brody's (1991):

(11)

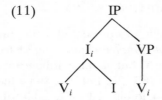

Consider English. The theory automatically accounts for the fact that V
cannot be moved to I. Movement of V to I creates the structure $[_I V + I]$,
in which the index of V cannot percolate to I, because I already has an
index. Thus I does not bind the trace of V, and the trace of V remains
unlicenced. However a full account of English must also explain why I is
allowed to move to V. The problem of course is that movement of I to V
yields the structure in (12), where the trace of I is not bound. What we can
assume is that the fact that I remains unlicenced as a trace is irrelevant to
the extent that the (I, V) dependency, with V unmoved, is licenced as an
expletive dependency:

(12)

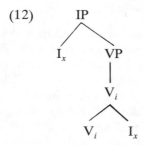

In other words, the situation in Italian and French can be summarized as
in (13), the situation in English as in (14). All of the configurations in
(13)–(14) are correctly accounted for under our theory. In English the
configuration in (14b) is obligatory if I is affixal, for reasons of morphological welformedness if we follow Chomsky (1989) as before; the configuration in (14c) is possible and obligatory if I is non-affixal, as in (15). If
what precedes is on the right track then, what we have shown is that the

strong/weak parameter in (4) can indeed be translated into a lexical/non-lexical parameter:

(13)   (French/Italian)
    a. $I \ldots V_i$

    b. *$I \ldots V_i$

    c. *$I \ldots V_i$

(14)   (English)
    a. *$I_x \ldots V_i$

    b. $I_x \ldots V_i$

    c. $I_x \ldots V_i$

(15)   John would completely lose his mind

Unfortunately, new patterns arise when auxiliaries are considered. In Italian *completamente* always follows and never precedes an auxiliary, exactly as a main verb. Similarly in French *complètement* always follows and never precedes auxiliaries in finite sentences, exactly as main verbs again. But in English *completely* follows and does not precede finite auxiliaries, contrary to main verbs. This state of affairs is summarized in (16)–(17):

(16)   En.   a. John has completely lost his mind
       Fr.   b. Jean a complètement perdu la tete
       It.   c. Gianni ha completamente perso la testa

(17)   En.   a. *John completely has lost his mind
       Fr.   b. *Jean complètement a perdu la tete
       It.   c. *Gianni completamente a perduto la testa

Since in French and in Italian the situation with auxiliaries is identical to that with main verbs, the same account will hold for (16)–(17) as for (8)–(9). For English however, it is obvious that some property of auxiliaries will induce a different behavior from that of main verbs.

In what precedes, we have derived the fact that a main verb cannot move to I in English from the assumption that I has an index of its own. The same conflict potentially arises between the index of I and the index of an auxiliary; but suppose the auxiliary itself is a dependent lexical head, technically associated with a variable index. If the auxiliary moves to I, an (expletive, trace) sequence is effectively created, hence a configuration independently known to be wellformed. Thus we account for why V-to-I is permissible with auxiliaries in English (16)—(17). As for why I-to-V is not permissible, Chomsky's (1989) Least Effort Principle can again provide the answer.

In summary, assuming that French and Italian auxiliaries are associated with a variable index like their English counterparts, we are proposing that the situation in (18) holds in French or Italian for auxiliary verbs, parallel to the situation in (13) for main verbs; while the situation in (19) holds for auxiliaries in English, at variance with the situation in (14) for main verbs. The movement configuration in (18a) and (19a) is forced if I is affixal; the non-movement configuration in (18c) is allowed and indeed forced in English if I is non-affixal, as in (20):

(18)   (French/Italian)
    a. $I \ldots V_x$

    b. $*I \ldots V_x$

    c. $*I \ldots V_x$

(19)   (English)
    a. $I_x \ldots V_x$

    b. $*I_x \ldots V_x$

    c. $I_x \ldots V_x$

(20)   John would completely have lost his mind

The paradigm exemplified in (8)—(9) and in (16)—(17) is only one of three main paradigms presented in Pollock (1989). The second paradigm concerns the distribution of floating quantifiers of the type of *all/tous/tutti*; this appears to simply repeat that of adverbs of the type of *completely/ complètement/completamente*. The third paradigm concerns the distribution of sentential negations. This appears to largely overlap with that of adverbs and floating quantifiers; since however the overlap is not complete, a separate examination of the paradigm is in order.

Consider English. In main sentences *not*, or its clitic counterpart *n't*, patterns with adverbs with respect to auxiliaries since it must follow them, but a problem emerges with main verbs. As expected, the negation cannot follow them, but at the same time it cannot precede them either unless *do* is inserted in I position. The entire paradigm is provided in (21)—(22). Thus the problem for English is what blocks (21b) and forces the adoption of (21c):

(21) a. *John left not (n't)
    b. *John not left
    c. John did not (n't) leave

(22) a. John has not (n't) left
    b. *John not has left
    c. *John does not (n't) have left

Suppose that following Pollock (1989), Chomsky (1989), Kayne (1989) the negation is itself taken to be a head generated between the I and the V heads. A decision is necessary within our theory as to whether the negation head is lexical or non-lexical. Suppose we take it to be lexical, and indeed associated with an non-variable index owing to its LF content. We can also assume that both *not* and *n't* represent the negation head in English, in its non-clitic and clitic form, respectively.

Consider first why I cannot move to V, yielding (21b). If I moves to V across the negation, a sequence including I and V violates locality, i.e., descriptively the Head Movement Constraint of Travis (1984). On the other hand a sequence including the negation as well as V violates the basic principle of compatibility of indices within a sequence, given that V and the negation have incompatible indices. If this explains the illformedness of (21b), since movement of V to I as in (21a) is independently excluded as before, then the choice of (21c) is forced. I must of course be in the *do* support form for reasons of morphological wellformedness.

The question, given that I movement to V is impossible across the negation, is how auxiliary movement to I is possible across it, as in (22a). Of course a sequence that involves only I and the auxiliary and bypasses the negation is illformed on locality grounds. However, given that the auxiliary has a variable index, a sequence can be formed, including I, the negation and the auxiliary on compatibility of indices basis, that also respects locality. The question why movement of I to the auxiliary is blocked, as in (22b) can be answered as usual by the Least Effort Principle.

As for the question why *do* cannot cooccur with an auxiliary, as in (22c), what we must have recourse to is again as in Chomsky (1989) some functional principle. Suppose I in English has two allomorphs, an affixal one and non-affixal *do*. We can assume that of the two allomorphs the affixal one is the default one. Thus *do* is resorted to only if the affixal form is blocked. Since the affixal form is not blocked in the presence of an auxiliary and negation, this means that *do* only surfaces with main verbs and negation.

Thus the situation with main verbs and negation in English can be schematized as in (23), the situation with auxiliaries and negation as in (24). The non-movement configuration in (23c) corresponds to the *do* support one; the non-movement configuration in (24c) though not possible with *do* is of course not impossible with I in general, as for instance in (25):

(23) a. $^*\mathrm{I}_x \ldots \mathrm{NEG}_i \ldots \mathrm{V}_j$

b. $^*\mathrm{I}_x \ldots \mathrm{NEG}_i \ldots \mathrm{V}_j$

c. $\mathrm{I}_x \ldots \mathrm{NEG}_i \ldots \mathrm{V}_j$

(24) a. $\mathrm{I}_x \ldots \mathrm{NEG}_i \ldots \mathrm{V}_x$

b. $^*\mathrm{I}_x \ldots \mathrm{NEG}_i \ldots \mathrm{V}_x$

c. $\mathrm{I}_x \ldots \mathrm{NEG}_x \ldots \mathrm{V}_x$

(25)   John would not/n't have left

By comparison with English, the negation system of Italian is straightforward. The clitic *non* appears in front of the V moved to I; a negative adverb of the type of *mica* (at all) appears after both auxiliaries and main verbs, exactly as other adverbs we have examined. The relevant paradigm then is as in (26)—(27), which translate the (a) examples in (21)—(22). All of these data are compatible with the negation being a head in Italian, if the negation is non-lexical in Italian. Thus the negation is just a head through which V moves on its way to I, picking up *non*. The position of *mica* or similar elements is also straightforwardly explained, on the assumption that they fill an adverb position:

(26)   Gianni non partí (mica)

(27)   Gianni non é (mica) partito

The final remaining case is French. In French again a clitic *ne* precedes V; but a correlate to *ne* such as *pas* is obligatory in French, contrary to

Italian where it is merely possible. The position of *pas* generally overlaps with that of adverbs. Thus *pas* follows main verb and auxiliary in finite sentences. The relevant paradigm is given in (28)—(29), which translate the (a) and (b) examples in (21)—(22). If *ne* is a head in French the examples in (28)—(29) are explained on the basis of it being a non-lexical head; V then simply moves through the negation on its way to I:

(28) a. Jean ne partit pas
     b. *Jean ne pas partit

(29) a. Jean n'est pas parti
     b. *Jean ne pas est parti

Thus the resumptive table for the interaction of negation and main verbs/auxiliaries is as in (30)—(31) for French or Italian. It is clear from (30)—(31) when compared to (23)—(24) that a parameter concerning negation also divides English from Italian/French. In particular *not* is lexical in English, while *ne/non* are nonlexical in Italian and French. This leaves the possibility open of treating *pas* as lexical and indeed bearing the LF content of negation in French, thus capturing the parallelism between *not* and *pas* argued for in Pollock (1989); and the same holds for the optional correlate of *non* in Italian. Independent evidence is needed to establish this parameter, but we leave this issue open for further research:

(30) a. $\text{I} \ldots \text{NEG} \ldots \text{V}_i$

    b. *$\text{I} \ldots \text{NEG} \ldots \text{V}_i$

    c. *$\text{I} \ldots \text{NEG} \ldots \text{V}_i$

(31) a. $\text{I} \ldots \text{NEG} \ldots \text{V}_x$

    b. *$\text{I} \ldots \text{NEG} \ldots \text{V}_x$

    c. *$\text{I} \ldots \text{NEG} \ldots \text{V}_x$

In conclusion, what we are proposing is extremely simple. Lack of movement to a I head is taken to correspond to the head acting as an expletive. This is true of English finite sentences, and appears by and large to extend to both English and French non-finite ones. By contrast, movement to a I head is taken to correspond to the head being marked for non-lexicality.

This characterizes French and Italian finite sentences as well as Italian non-finite ones. Germanic languages appear to distribute themselves over the two groups, mainland Scandinavian patterning with English, and Icelandic or German/Dutch with French or Italian. If our theory is on the right lines, we expect that V-to-C behaviors should also be reducible (in part) to the lexical/non-lexical parameter. Again this prediction must be left open for further research.

Finally, the parameter in (10) does correlate with a different parameter, a non-syntactic one, namely whether a given head is affixal or not. Roughly, languages which in our terms have non-lexical I are characterized by affixal I, as in Italian and French, respectively. Languages which in our terms have lexical I are characterized by non-affixal I, such as English modals, or zero inflection, such as mainland Scandinavian. However the correlation is not perfect; languages such as English can have affixal I, at a price, the price being lowering and *do* support. Thus the two parameters appear to be irreducible. This of course predicts that there will be affixes that are lexical. Indeed we can take N-incorporation to be prima facie evidence to support this claim.

In the next section, we will discuss how two languages like French and Italian that are both characterized by movement of V to I differ in the properties this movement has; and we will argue that this difference corresponds to the null subject parameter.

3. THE NULL SUBJECT PARAMETER

While French and Italian pattern against English with respect to V-to-I, French and English, two typical non null subject languages pattern against Italian, a typical null subject language, with respect to the null subject phenomena in Rizzi (1982). All three languages have canonical lexical subjects, as in (32). But Italian has, and French and English do not have null subjects, as in (33); Italian has, and French and English do not have inverted lexical subjects, as in (34)—(36); and finally French and English have, and Italian does not have, C-t and *wh*-t effects, as in (37)—(38):

(32)   It.   Gianni ama Maria
        Fr.   Jean aime Marie
        En.   John loves Mary

(33)   It.   Ama Maria
        Fr.   *Aime Marie
        En.   *Loves Mary

(34)   It.   É arrivato un bambino
        Fr.   *Est arrivé un enfant
        En.   *Arrived a child

(35)   It.   Ha telefonato un bambino
        Fr.   *A telephoné un enfant
        En.   *Phoned a child

(36)   It.   Ha scritto una lettera un bambino
        Fr.   *A écrit une lettre un enfant
        En.   *Has written a letter a child

(37)   It.   Chi pensi che sia venuto
        Fr.   *Qui penses-tu que est venu
        En.   *Who do you think that came

(38)   It.   Chi si chiede perché é venuto
        Fr.   *Qui se demande-t-il pourquoi est venu
        En.   *Who does he wonder why came

Our idea is that all of the null subject data are accounted for if V moves to I in a null subject language like Italian and adjoins to it; but the derived V + I structure is a V, not an I, as we have assumed so far. Thus the canonical S-structure of an Italian sentence will include the subtree in (39), not the one in (11):

(39)

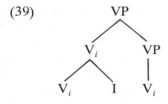

By contrast, if our idea is correct, a non-null subject language such as French or English is characterized either by the verb not moving to I or by it moving to I. If movement takes place, as in French, the relevant structure is as in (11). If no movement takes place, on the other hand, or downward movement of I to V does, as in English, a structure of the type in (12) is obtained. What we must now show is that the distinction between (39) and (11) or (12) is the relevant one for the null subject parameter. If we are correct, at least all of the data in (32)—(38) should be derivable from it.

Under the Lexical Parameterization Principle, the null subject parameter itself can reduce simply to whether I is or is not a head, as in (40), a suggestion put forward in Brody (1989). If it is a property of I that it is a head, then this property is satisfied in (11) before and after V movement to I. The same is true in the absence of V movement to I, as in (12). In (39), however, I is not a head, since it does not project in the X-bar theoretical sense. Thus the structure in (11) or (12) is forced. If on the other hand I is not a head, the structures in (11) and (12) are blocked, since the IP projection is not licenced under X-bar theory. Thus the structure in (39) is forced:

(40)   *(Null Subject) Parameter*
        α is/is not a head

Intuitively, the content of the parameter in (40) is that null and non-null subject languages differ along the synthetic/analytic divide. Null subject languages are inflectionally synthetic languages where I is totally incorporated into V. Non null subject languages are analytic languages, where I is only partially or not at all incorporated into V. In terms of the theory we are putting forward, English I, having the status of a lexical head, can then be construed as stronger than French I, which has the status of a non-lexical head; the latter in turn can be construed as stronger than Italian I which is not a head. This intuitive hierarchy of strength reverses Chomsky's (1989) ranking of French (and Italian) I as strong with respect to English weak I. Similarly it reverses Kayne's (1989) ranking of Italian I as strong with respect to French (and English) weak I.

There is an obvious reason for this apparent contradiction. Chomsky's (1989) or Kayne's (1989) hierarchies are based on considerations of morphological strength; ours on considerations of syntactic strength. In the languages under exam these two scales of strength appear to run exactly opposite to one another. However our theory makes no (direct) prediction as to the morphological realization of I in the various types of languages. This lack of predictions receives prima facie support by the fact that null subjects are found in the absence of (overt) inflection in languages like Chinese; the same fact forces the claim that morphological homogeneity, as opposed to morphological strength, is relevant for the null subject parameter, as in Jaeggli and Safir (1989).

Consider now the null subject phenomenon, as in (33). Suppose that under the Extended Projection Principle I projects a subject position in (39). What we want to suggest is that an empty category in subject position is licenced as an argument because I is in a clitic configuration with respect to V. Suppose on the other hand I projects an empty position in French where the relevant configuration is as in (11). The empty position cannot be licenced simply because I is a head. Similarly, consider English, as in (12). The licencing of an empty category subject is made impossible in (12), exactly as in (11), by the fact that I is a head and is not a clitic on a head. Thus the subject must be lexical.

On the other hand, lexical subjects, as in (32), are possible not only in languages like English and French, but also in languages like Italian. In Italian and other languages where I is not a head, what we must assume is that the presence of lexical subjects is licenced as a case of clitic doubling. Motivating this assumption implies showing that Italian independently has clitic doubling, and that clitic doubling occurs without insertion of an extra Case-marker in front of the doubled DP. There is indeed a prima-facie candidate for clitic-doubling without P insertion in Italian, namely Clitic Left (and Right) Dislocation, in the sense of Cinque (1991).

The derivation of inversion sentences as in (34)—(36) is comparatively simple. The standard theory holds as in Rizzi (1982) and Burzio (1986) that the possibility of inversion depends on the possibility of a null subject expletive; thus inverted subjects must be related to an expletive, and this can be empty in Italian. This theory straightforwardly transposes to our present framework, since we have already seen that an empty subject can be licenced in (39), while in (11)—(12) only a lexical subject is.

Finally we are in a position to explain the properties of subjects under extraction, as in (37)—(38). We follow Rizzi (1990) in assuming that a subject trace in English or French cannot be immediately adjacent to a C as in (37) or to a wh-phrase in the Spec of CP, as in (38), because though it is governed by I, it is not c-commanded by it. It is wellknown that while English (37) can be rendered grammatical by the deletion of *that*, French (37) can be rendered grammatical by an apparent agreement of *que* with I, to give *qui*, as in (41). This suggests, as in Rizzi (1990), that the zero C in English is simply the form of C that agrees with I, as *qui* is in French and that the reason for having an agreeing form in C, is that the trace can then be governed and c-commanded by C. Similarly the reason for the impossibility of a *wh*-phrase in the Spec of CP is that in this case Spec-head agreement between C and the *wh*-phrase preempts agreement of C with I:

(41)   En.   Who do you think came
        Fr.   Qui penses-tu qui est venu

If French and English (37) and (38) are excluded on c-command grounds, extraction of the subject can be wellformed in (37) and (38) in Italian, simply because inversion is available in the language, as indeed proposed in Rizzi (1982). Thus, subject extraction need not take place from the Spec of IP position, but can take place from an inverted position, which is c-commanded by its governor, I. It follows that our parameter automatically derives the extraction facts in (37)—(38) by deriving the inversion facts.

Our proposals are easily summed up. Categories differ with respect to whether they are lexical or non-lexical, and with respect to whether they are heads or not in the X-bar theoretical sense. In particular, if I is lexical, it functions as an expletive with respect to V; if I is non-lexical, V moves to it. In other words, the +/− lexical parameter represents the V-to-I parameter. On the other hand, if I is a head, movement of V to I, when possible creates an adjoined structure $[_I V + I]$, with an IP projection; if I is not a head, movement of V to I creates an adjoined structure $[_V V + I]$, with a VP projection. If we are correct, the structure resulting in the former case is to be identified with a non null subject structure, the structure resulting in the latter case with a null subject structure.

There is at least one important theoretical issue that our theory contributes to. D-structure is standardly conceived as the level of representation prior to the application of movement, or viceversa the level of representation derived by the undoing of movement. If so, the question arises what the D-structure of (39) looks like before V moves to I. The easiest solution is that the D-structure counterpart to (39) is as in (42); in (42), then, [e]P represents a projection of the empty head e, whose value is fixed at V after movement:

(42)

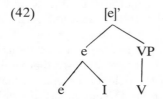

NOTE

* The initial version of this article, circulated at the Knowledge and Language Conference can be found in Manzini (1989). The present version reflects in part work carried out at NIAS, Wassenaar in the fall/winter of 1990/91 as part of the research project on "The Logical Problem of Language Acquisition" coordinated by T. Hoekstra and H. van der Hulst.

REFERENCES

Borer, Hagit: 1984, *Parametric Syntax*, Foris, Dordrecht.
Brody, Michael: 1989, 'Old English Impersonals', in R. Carston (ed.), *UCL Working Papers in Linguistics* 1, 262—294.
Brody, Michael: 1990, 'Economy, Earliness and LF Base Syntax', in H. van de Koot (ed.), *UCL Working Papers in Linguistics* 3, 25—32.
Burzio, Luigi: 1986, *Italian Syntax*, Reidel, Dordrecht.
Chomsky, Noam: 1981, *Lectures on Government and Binding*, Foris, Dordrecht.
Chomsky, Noam: 1982, *Some Concepts and Consequences in the Theory of Government and Binding*, MIT Press, Cambridge, Massachusetts.
Chomsky, Noam: 1989, 'Some Notes on Economy of Derivation and Representation', in I. Laka and A. Mahajan (eds.), *MIT Working Papers in Linguistics* 10, 43—74.
Cinque, Guglielmo: 1990, *Types of A'-Dependencies*, MIT Press, Cambridge, Massachusetts.
Jaeggli, Osvaldo: 1982, *Topics in Romance Syntax*, Foris, Dordrecht.
Jaeggli, Osvaldo and Ken Safir: 1989, 'Introduction', in O. Jaeggli and K. Safir (eds.), *The Null Subject Parameter*, Kluwer, Dordrecht, 1—44.
Kayne, Richard: 1989, 'Null Subjects and Clitic Climbing', in O. Jaeggli and K. Safir (eds.), *The Null Subject Parameter*, Kluwer, Dordrecht, 239—261.

Manzini, M. Rita: 1989, 'Categories and Acquisition in the Parameters Perspective', in R. Carston (ed.), *UCL Working Papers in Linguistics* **1**, 181—191.

Manzini, M. Rita: to appear, *Locality*, MIT Press, Cambridge, Massachusetts.

Manzini, M. Rita and Kenneth Wexler: 1987, 'Binding Theory, Parameters and Acquisition', *Linguistic Inquiry* **18**, 413—444.

Pollock, Jean-Yves: 1989, 'Verb Movement, Universal Grammar and the Structure of IP', *Linguistic Inquiry* **20**, 365—424.

Rizzi, Luigi: 1982, *Issues in Italian Syntax*, Foris, Dordrecht.

Rizzi, Luigi: 1990, *Relativized Minimality*, MIT Press, Cambridge, Massachusetts.

Travis, Lisa: 1984, *Parameters and Effects of Word Order Variation*, Ph.D. dissertation, MIT, Cambridge, Massachusetts.

Wexler, Kenneth and M. Rita Manzini: 1987, 'Parameters and Acquisition in Binding Theory', in T. Roeper and E. Williams (eds.), *Parameter Setting*, Reidel, Dordrecht, 41—76.

CELIA JAKUBOWICZ

LINGUISTIC THEORY AND LANGUAGE ACQUISITION FACTS: REFORMULATION, MATURATION OR INVARIANCE OF BINDING PRINCIPLES

1. INTRODUCTION

Since the emergence of the Generative Standard Theory framework (cf. Chomsky 1965, 1981, 1986), language acquisition has been idealized as an instantaneous process resulting from the interaction of a small number of innately determined linguistic principles and the linguistic experience available to the child. According to this model, the new born baby is equiped with a set of computational mechanisms for manipulating structural representations that are ready to be used. Once the genetic device gets down to work, only certain adjustments, determined by the linguistic environment, are needed for the child to master the grammar of the language to which he is exposed. Under this approach, we expect children's linguistic behavior to precociously conform to the Principles of Universal Grammar (henceforth UG) or at least, we expect the transition from the initial to the steady state of language acquisition to be rapid and error free. However, such a prediction is apparently at odds with certain facts which indicate that many aspects of the acquisition of language knowledge are sequentially ordered.

To solve the problem raised by these facts, different solutions have been proposed, which are still a matter of current debate among those interested in a theory of the child's real-time growth of language. In this paper I discuss four different ways of dealing with the developmental problem. The four proposals are alike in that they assume that the Principles of UG are innately determined, and that the input data are not ordered. They are however different with respect to both the explanatory mechanisms they involve and the empirical predictions they give rise to. Throughout, I discuss the four solutions presented in (a) through (d) below with respect to the acquisition of overt anaphors and pronouns, whose linguistic behavior is explained by the Principles A and B of the Binding Theory. I assume the version of the Binding Theory put forward in Chomsky (1986), in which the relevant governing category for the expression α is the least complete functional complex containing a governor for α in which α could satisfy the Binding Theory with some indexing I. A complete functional complex consists of a lexical head and all grammatical functions compatible with that head (the complements and the subject). According to Principle A, if α is an anaphor, it must be

157

Eric Reuland and Werner Abraham (eds), Knowledge and Language, Volume I, From Orwell's Problem to Plato's Problem: 157—184.
© 1993 *by Kluwer Academic Publishers. Printed in the Netherlands.*

bound in its governing category; according to Principle B, if α is a pronominal, it must be free.

The acquisition of overt anaphors and pronouns has been studied in a variety of languages. Otsu (1981), Jakubowicz (1984, 1987), Wexler and Chien (1985), McDaniel, Cairns and Hsu (1987) and Solan (1987) tested young native English speaking children. Padilla-Rivera (1985) interviewed native Spanish speaking children. Deutsch, Koster and Koster (1986) provided data for Dutch, Jakubowicz and Olsen (1988) for Danish and Jakubowicz (1989a and b) for French. The authors used different experimental tasks, presented somewhat different linguistic material and tested either only young children and/or also older children. Such methodological differences may in part explain why the results obtained are heterogeneous with respect to the age at which children master both types of expressions. However, in spite of language specific differences with regard to the argumental and phrase structure status of the anaphoric and pronominal expressions on the one hand, and in spite of the incidence of the experimental factors just mentioned on the other hand, the results so far available show that there is an age at which children perform better with sentences containing an anaphor than with their counterpart with a pronoun. Leaving aside the fact that in certain studies the acquisition lag shows up in 3 to 4 years olds while in certain others it appears at later ages (5 to 6 years old children), we are confronted with the kind of developmental problem mentioned above, namely, the fact that at a certain point the child's performance with respect to linguistic facts accounted for by the Principle B of the Binding Theory is less perfect than his performance with regard to facts accounted for by the Principle A. In other words, the acquisition of certain aspects of the behavior of overt anaphors and pronouns appears to be sequentially ordered.

Let us now go through the four different solutions that have been proposed to explain this fact:

a) According to Grimshaw and Rosen (1990) the acquisition lag between anaphors and pronouns is in fact a consequence of the use of experimental paradigms "that necessarily underestimate children's command of Principle B and overestimate their command of Principle A" (p. 2). I will critically review this proposal and show that the experimental artifacts invoked by these authors are unable to explain both the French data I will present and most of those provided by previous research.

b) A different response to account for the acquisition lag between anaphors versus pronouns is to argue that Principle B of the Binding Theory should be reformulated, so that it would be concerned only by bound variable pronouns. Under this view, proposed by Wexler and Chien (1985, see also Chien and Wexler 1987, 1988, Montalbetti and Wexler 1985), children should perform better with respect to pronouns bound by a quantifier than with respct to non-variable pronouns. Further, at the age at which children show successful comprehension of sentences containing anaphors they should be equally successful with sentences containing a pronoun bound by a quantifier. Contrary to these predictions, our data will show that there is no difference in the comprehension of bound variable and non-variable pronouns, and that the acquisition lag between anaphors versus pronouns persists even if the pronoun is in an environment in which it must be interpreted as a bound variable. I will propose that the acquisition results do not justify the reformulation of Principle B.

c) Does the child's imperfect performance with regard to pronouns reflect violations of Principle B that would follow if that Principle were maturationally available later than Principle A as suggested by Felix (1984, 1988)? Do the data provide evidence for other maturational claims, for instance, for the claim that the capacity of "A-chain formation undergoes maturation", put forward by Borer and Wexler (1987, 1988)? To answer these questions I will suggest that children's successful comprehension with pronouns in subject position provides strong evidence in support of the availability of Principle B. Then, the explanation of the fact that the same children produce binding errors with regard to pronouns in object position, would need to appeal to factors other than the maturational one. I will also show that children as young as three successfully comprehend sentences with anaphoric clitics. According to Kayne (1988), in French, (Spanish and Italian) sentences with reflexive clitics involve NP movement, that is "A-chain formation". The finding that young French speaking children master sentences with anaphoric clitics (and full verbal passives) but misinterpret other constructions involving NP movement (i.e., raising), would make it difficult to accept the proposition that the capacity of A chain formation matures.

d) A fourth solution is provided by the invariance of UG Principles plus lexical learning hypothesis. According to this hypothesis children have access to the same set of UG Principles all along language development although their knowledge of properties of lexical (and morphological) items determines which Principles are operative at a given moment (cf. Jakubowicz, op. cit., Pinker 1984, Clahsen 1988). The results of the two experiments that are presented in the next section will allow us to conclude that the acquisition lag between anaphors and pronouns is best explained in terms of this hypothesis.

2. THE EXPERIMENTS

2.1. *The Interpretation and the Production of Sentences with Reflexive Anaphors and Pronouns*

2.1.1. *Methodology*

In this experiment I employed a sentence-picture matching task and an elicited production task.

Subjects were 104 native French speaking children between the ages of 3.0 and 7.5 years, equally divided in four age groups (Years. months: 3.0— 3.5, 3.6—4.0, 5.0—6.0 and 6.1—7.5). Mean ages for each group were: 3.4, 3.9, 5.8 and 6.11.

For the interpretation task, the stimuli consisted of 8 test sentences presented within a list containing a total number of 24 sentences. Four out of the 8 test sentences contained the third person clitic reflexive **se**, and four contained the third person clitic pronoun **le/la**. Sentences (1) and (2) are examples of test sentences that appeared in the experiment. The complete list of pre-test and test-sentences used is presented in the appendix.

(1)   Kiki dit   que Nounours **se** brosse
        Kiki says that Teddy Bear brushes himself

(2)   Schtroumpfette dit   que Barbie **la** brosse
        Smurf (female) says that Barbie brushes her

A standard sentence-picture matching task served as the task of interpretation. For each test-sentence as well as for each filler, subjects were required to choose, among three pictures, the one that best matched the sentence. The picture array was composed so that only one picture of the three was the correct choice. The other two pictures depicted interpretations of the test sentence that deviated in a particular way from the test sentence, and were thus incorrect choices. For both sentence-types, one of the incorrect choices showed a scene in which the actors were the correct ones in the test sentences, but the action (reflexive versus non reflexive) did not agree (i.e., pictures (b) in arrays 1 and 2 below). For test-sentences with a reflexive anaphor, a second incorrect choice was one of the following two: (i) the actors are those mentioned in the test-sentence but the reflexive is bound to the matrix subject (picture (c) in array 1); (ii) the actor and the action described in the embedded clause are correct, but the character mentioned in the matrix does not agree with the one presented in the picture. With respect to sentences containing a pronoun, the second incorrect choice was one of the following two: (i) the action is the correct one in the test sentence but the relation between the actors is reversed; (ii) the actor and the action described in the embedded clause are correct but

Nounours dit que Kiki se lave
Teddy Bear says that Kiki washes himself

Picture array 1.

Nounours dit que Kiki le peigne
Teddy Bear says that Kiki combs him

Picture array 2.

162   CELIA JAKUBOWICZ

the character mentioned in the matrix of the test sentence is changed to a
different sex character (picture (c) in array 2).

As illustrated in the picture arrays, three positions were determined on
each presentation page, so that a correct choice could appear in each one
of these positions. Occurrences of picture types (corresponding to correct
and incorrect choices) were the same for each one of the three positions
on the page.

Two types of questions, illustrated in (3) and (4) were presented to
elicit from the child the reflexive **se** and the pronoun **le/la**. Each type of
question was repeated 8 times.

(3)   Que fait Nounours?   (The experimenter points
       What is Teddy Bear doing?   to scene (c) in picture
                                        array (1) above)

(4)   Que fait Kiki à Nounours?   (pointing to scene (b) in
       What is Kiki doing to Teddy Bear? picture array (1))

2.1.2. *Results*

i. *The Sentence-picture Matching Task.* Consider Figure 1 where I report
the percentages of correct responses for overt anaphors and pronouns as a
function of age.

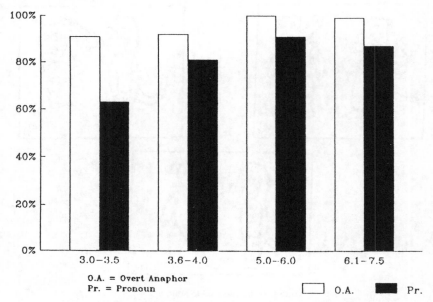

O.A. = Overt Anaphor
Pr. = Pronoun   ☐ O.A.   ■ Pr.

Fig. 1. Percentage of correct responses for overt anaphors and pronouns × age.

Sentences with the reflexive **se** were correctly interpreted by all the children whereas for the pronoun **le/la** errors were made at all ages. However both the frequency and the nature of the errors were different according to the children's age. First, the younger children of the sample — the 3.0–3.5 year olds — gave correct responses for reflexives significantly more often than for pronouns ($F[1, 25] = 30.34$; $p < 0.0005$) and showed a relatively important number of binding errors with regard to pronouns (i.e., they pointed to (b) in picture array 2 above). Second, for the remaining age groups there was also an effect of expression type: that is, correct responses for reflexives were significantly more frequent than for pronouns ($F[1, 25] = 6.16$; $p < 0.02$ for Age 2; $F[1, 25] = 7.88$; $p < 0.02$ for Age 3 and $F[1, 25] = 7.50$; $p < 0.02$ for Age 4). Contrary to the younger ones, children older than 3.5 performed few binding errors; these children almost always interpreted the pronoun as free but they sometimes reversed the subject-object relations encoded in the test sentences and thus pointed to an incorrect picture. The fact that these grammatical relation errors arose only when the children were presented with arrays in which the three pictures contained the same two characters suggests that these errors were due to a performance factor rather than to the child's linguistic knowledge of the behavior of pronouns. Further analysis revealed that the difference in the percentage of correct responses between the first two age groups was significant ($F[1, 50] = 12.91$; $p < 0.001$) while the difference between the remaining age groups was not significant. This result indicates that there is a developmental change in the interpretation of sentences with pronouns, and that this change occurs primarily between the mean ages of 3.4 and 3.9.

ii. *The Elicited Production Task.* Consider first Figure 2 in which each whole column indicates the percentage of sentences with a clitic or a strong postverbal pronoun produced by the children of each age group in response to question-type (3) for a picture representing a self-oriented action. Grammatically correct scores are shown in the white portion of each column, incorrect scores in the portion with stars.

Almost all grammatically correct productions were of the form: **il/elle se V** (i.e., "il se lave" — 'he SELF washes'); sentences of the form **il/elle se V NP** (i.e., "il se lave le/son visage" — 'he SELF washes the/his face') were also observed, but considerably less often than the former. As seen, the score for grammatically correct sentences was high and uniform across ages; the difference in the number of correct answers between ages was not significant. Only 30 out of 664 sentences (data collapsed over age) were incorrect; in 26 cases a gender error in subject position, **i** or **il** instead of **elle** ('he/she') gave rise to incorrectness. Four different children produced, each child once, sentences such as (5) in which the postverbal pronoun was used deictically (the child pointed to the doll in

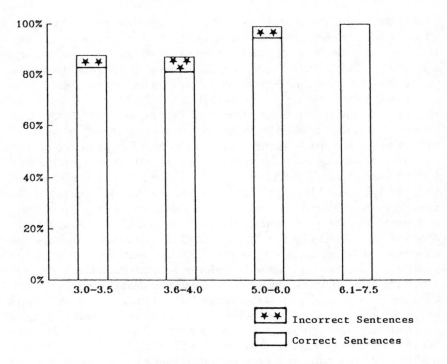

Fig. 2. Elicited production for reflexive actions × age.

the picture). As known, for French a strong pronoun would not normally appear in object position.

(5)   *"elle lave elle"   (where Smurf (female) washes herself)
        she washes she

Consider now Figure 3 in which I present the scores for grammatically correct sentences and for incorrect sentences for question type (4)/picture representing a non-self-oriented action.

The most frequent grammatically correct productions were of the form: **il/elle le/la V** (i.e., "il le brosse" — 'he him brushes'). Forms such as **il/elle lui V NP** (i.e., "il lui brosse les cheveux" — 'he to him brushes the hairs') were also produced, but less often than the former. Sentences with errors are discussed below. As we can see from Figure 3, while the amount of sentences with a clitic or a strong postverbal pronoun is important even at the youngest age group (70% of productions), 3—3.5 year olds produced grammatically correct sentences as often as incorrect ones. The amount of incorrect productions decreases with age but only the 6—7.5 year olds showed 100% grammatically correct sentences. The major types of children's errors in the production of sentences with pronouns are

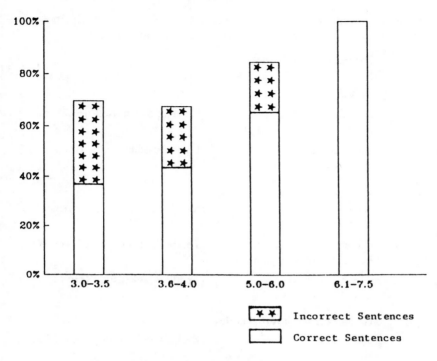

Fig. 3. Elicited production for non reflexive actions × age.

presented in (6). The percentage of errors is calculated in relation to the total number of incorrect responses observed at each age group (65 at the first age group; 47 at the second age group and 42 at the third).

(6)   **Distribution of errors in the production of sentences with pronouns. Percent of errors in each age group**

Error-type	3.0—3.5	3.6—4.0	5.0—6.0	6.0—7.5
Expression: *se* instead of *le/la*	34.4	21.3	4.9	0
Case: *lui/le* instead of *le/lui*	25.0	31.9	36.6	0
Gender (clitic): *le/la*	21.8	36.2	36.6	0
Gender (subject): *il/elle*	18.7	10.6	21.9	0

Examples of sentences with errors are given below. A description of the scene presented to the child and the child's age are in brackets.

(7)   **Expression-type errors**

   a.   *i se lave (points to Teddy Bear) (Kiki washes Teddy Bear) (3.5)
          he SELF washes

   b.   *il se coiffe Schtroumpfette   (Kiki combs Smurfette) (3.2)
          he SELF combs Smurfette

   c.   *elle se coiffe lui   (Barbie brushes Teddy Bear) (3.2)
          she SELF combs him

(8)   **Case errors**

   a.   *elle lui essuie   (Smurfette washes Barbie) (3.6)
          she to her dries

   b.   *il le brosse ses cheveux   (Teddy Bear brushes Kiki) (3.4)
          he him brushes the hairs

   c.   *il lui brosse Kiki   (Teddy Bear brushes Kiki) (3.3)
          he to him brushes Kiki

(9)   **Gender errors in verbal clitics**

   a.   *elle la coiffe   (Barbie combs Teddy Bear) (3.0)
          she her combs

   b.   *elle le peigne   (Barbie combs Smurfette) (3.4)
          she him combs

(10)   **Gender errors in subject pronouns**

   a.   *elle le brosse   (Teddy Bear brushes Kiki) (3.0)
          she him brushes

   b.   *il la mouche   (Smurfette wipes Barbie's nose) (3.5)
          he her wipes

The analysis of the production data obtained in this study detected children who overgeneralized **se**, children who allowed free variation among the available items and children for whom correct usage was established (both with respect to expression type and agreement features). Interestingly, the few children who overgeneralized **se** in the production task committed binding errors in the sentence-picture matching task. However, children who did not show binding errors in this task occasionally produced one or another error type in the elicited production task. This last fact suggests that these children were able to correctly identify the proform available in the input while their performance was less accurate when they were required to find the proform by themselves in the production task. Indeed, the 3 to 6 year olds gave significantly more correct responses for pronouns in the sentence-picture matching task than in the elicited production task (t[25] = 2.294; p < 0.05 at Age 1; t[25] = 3.539; p < 0.01 at Age 2 and t[25] = 3.308; p < 0.01 at Age 3), while for reflexives, the number of correct answers was equally frequent in the two tasks. (In the discussion section we come back to the facts here reported.)

2.2. *The Interpretation of Reflexive Anaphors, Variable and Non-variable Pronouns*

2.2.1. *Methodology*

In this experiment I employed an act-out task. Subjects were 80 native French speaking children equally divided in 4 age groups: 3.1—3.6; 3.7—4.0; 4.1—4.6 and 4.7—5.0 (years.months). Mean ages for each group were: 3.4; 3.9; 4.3 and 4.8.

The stimuli consisted of (i) sentences with a variable pronoun in object position, as in (11), or in subject position as in (12).

(11)   **Chaque** Schtroumpfette veut que Marie **la** brosse
        Every Smurf (female) wants that Marie brushes her

(12)   **Chaque** Barbie dit qu'**elle** est fatiguée
        Every Barbie says that she is tired

(ii) Sentences with a non-variable pronoun in object position as in (13), or in subject position as in (14).

(13)   Blanche Neige veut que Marie **la** mouche
        Snow White wants that Marie wipes her (nose)

(14)   Chaperon Rouge dit qu'**elle** est contente
        Little Red Riding Hood says that she is happy

(iii) Sentences with a reflexive anaphor as in (15).

(15)   Chaperon Rouge veut que Marie **se** peigne
        Little Red Riding Hood wants that Marie combs herself

Each test-sentence (and filler) was repeated 4 times. For sentences with the quantifier "chaque", children were presented half of the time with 2 dolls and half of the time with 3 dolls. Children were presented with 12 pretest-sentences. Only the children who succeeded in the pretest session were included in the sample. Examples of pretest-sentences are given below. The entire list of pretest and test-sentences is presented in the appendix.

168 CELIA JAKUBOWICZ

(16)   Pretest-sentences (examples)

a. **Chaque** Barbie saute
Every Barbie jumps up and down

b. **Chaque** Kiki peigne Chaperon Rouge
Every Kiki combs Little Red Riding Hood

c. Blanche Neige lave **chaque** Nounours
Snow White washes every Teddy Bear

d. **Chaque** Schtroumpfette est endormie
Every Smurf (fem.) is asleep

2.2.2. *Results*

Consider first the percentages of correct responses for variable and non-variable pronouns in subject position as a function of age. These data are represented in Figure 4. As seen, the frequency of correct responses is high for both variable and non-variable pronouns. However, the 3.1 to 3.6 year olds gave correct responses for non-variable pronouns significantly more often than for variable pronouns ($F[1, 19] = 5.44$; $p < 0.02$), while the difference in the amount of correct responses between these two expressions was not significant for the remaining age groups.

Consider now the Figure 5, where I present the percentages of correct responses for non-variable and variable pronouns in object position as a function of age. Children in each group performed slightly better with

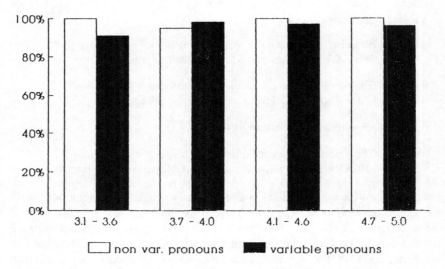

Fig. 4. Non variable and variable pronouns in subject position. Percentage of correct responses × age.

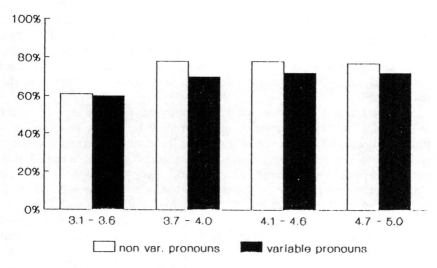

Fig. 5. Non variable and variable pronouns in object position. Percentage of correct responses × age.

non-variable pronouns than with variable pronouns. However, the difference in the amount correct between the two proforms was not significant within any of the age groups.

Further analysis revealed that children within each age group were significantly more successful with both non-variable and variable pronouns in subject position than with both non-variable and variable pronouns in object position (for non-variable pronouns $F[1, 19] = 12.70$; $p < 0.005$ at Age 1; $F[1, 19] = 4.0$; $p < 0.05$ at Age 2; $F [1, 19] = 7.12$; $p < 0.01$ at Age 3 and $F[1, 19] = 8.60$; $p < 0.01$ at Age 4; for variable pronouns: $F[1, 19] = 10.23$; $p < 0.005$ at Age 1; $F[1, 19] = 8.56$; $p < 0.01$ at Age 2; $F[1, 19] = 8.64$; $p < 0.01$ at Age 3 and $F[1, 19] = 6.73$; $p < 0.02$ at Age 4).

The role played by the syntactic position in which the proform appears also shows up in the analysis of individual patterns of responses. While for pronouns in subject position 67 out of 80 children responded equally well for both variable and non-variable pronouns, for pronouns in object position only 40 out of 80 children responded equally well to variable and non-variable pronouns. Further, the analysis of individual patterns revealed that when the pronoun appears in subject position 11 children gave more correct responses for non-variable pronouns than for variable pronouns, while only 2 showed the opposite pattern. When the expressions appear in object position, 23 children gave more correct responses for non-variable pronouns than for variable pronouns while 17 gave more correct responses for variable than for non-variable pronouns.

Binding errors were performed only with respect to pronouns in object position. The frequency of these errors decreases with age but binding errors were performed with respect to both sentences with non-variable pronouns and sentences with variable pronouns. Moreover only the younger children of the sample perform a few errors with respect to the quantifier.

Consider finally Figure 6 where I report the percentages of correct answers for reflexive anaphors and pronouns in object position as a function of age. As seen for the clitic **se** the frequency of correct responses is high and uniform across ages. The statistical analysis revealed that correct responses were significantly more frequent for anaphors than for variable pronouns (t[19] = 3.334; p < 0.01 at Age 1; t[19] = 2.703; p < 0.02 at Age 2; t[19] = 2.987; p < 0.01 at Age 3 and t[19] = 2.544; p < 0.02 at Age 4), or non-variable pronouns (t[19] = 3.035; p < 0.01 at Age 1; t[19] = 2.21; p < 0.05 at Age 2; t[19] = 2.706; p < 0.02 at Age 3 and t[19] = 2.881; p < 0.02 at Age 4).

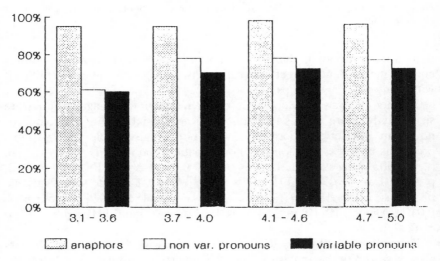

Fig. 6. Percentage of correct responses for anaphors and pronouns in object position × age.

2.3. *Summary of the Two Experiments*

The results of the two experiments above can be summarized as follows: children as young as 3 correctly interpret sentences with a reflexive and sentences with a non-variable or with a variable pronoun in subject position (i.e., Figure 7). With respect to sentences with a non-variable pronoun in object position, children perform better in the sentence picture

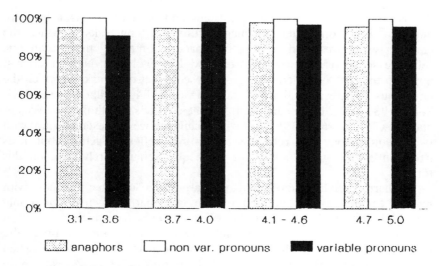

Fig. 7. Percentage of correct responses for anaphors and pronouns in subject position × age.

matching task than in the act-out task. However, binding errors with regard to pronouns show up in both tasks. As for variable and non-variable pronouns in object position the analysis of individual patterns of responses reveals that the number of children who perform better with non-variable pronouns is higher than the number of children who perform better with variable pronouns. However, at a group level, the difference in the amount of correct responses between the two proforms is not significant. In the act-out task binding errors are performed with respect to both non-variable and variable pronouns in object position. Thus, children show successful comprehension of anaphors and pronouns in subject position, but their performance is less perfect with regard to both non-variable and variable pronouns in object position.

3. DISCUSSION

Let us now take up the four explanations brought out in the introduction.

(a) Does the observed acquisition lag between anaphors versus pronouns in object position arise from certain experimental artifacts?

Grimshaw and Rosen (op. cit.) suggest that children's success with respect to anaphors is inflated by a strategy of choosing "reflexive action" pictures. If such a strategy were applied, since in our first experiment every picture array contained two reflexive-action picture choices, we would expect the children to perform at chance. As the results indicate, this is not the case. Note that Grimshaw and Rosen's criticism does not apply to data from other sentence-picture matching tasks in which the children were presented with picture arrays containing more than one reflexive scene (cf. Deutsch, Koster and Koster 1986). Neither does it apply to the data obtained through the use of "act-out" tasks in which the child must create an action (cf. Jakubowicz 1984, Padilla Rivera 1985, Solan 1987, etc.). It appears then that their criticism is limited to Wexler and Yu-Chien's study (1985), in which children were presented only with two picture-choices: a reflexive and a non-reflexive action, but it is irrelevant with regard to most of the acquisition research done on this topic.

Furthermore, Grimshaw and Rosen argue that children's success with respect to anaphors is inflated by the fact that the verbs commonly employed in stimuli belong to the class of verbs for which there is an intransitive use with a reflexive interpretation (i.e., **She washed, She dressed**). Under this assumption, we would expect children to fail when presented with verbs that do not belong to this class. As shown in Jakubowicz and Olsen (1988) for Danish, and in Jakubowicz (1989b) for French, 3 year olds are highly successful with anaphors when presented with verbs as **shine, shower, point, aim, serve, pour, prepare,** for which there is no intransitive use with a reflexive interpretation. Thus, contrary to Grimshaw and Rosen's suggestion, the choice of verbs in the experiments does not seem to account for the asymmetry in performance with sentences containing reflexives and pronouns in object position.

Grimshaw and Rosen also argue that children's imperfect performance on test-sentences with a pronoun is due to the presentation of sentences such as (17) that fail to provide a prominent discourse antecedent for the pronoun.

(17)   The Smurf is talking to him
         Grimshaw and Rosen, example (11)

Children's failure with sentences like (17) will not reflect lack of knowledge of the Binding Theory; rather it will be a consequence of the child's difficulty to construct a sensible interpretation of a sentence containing an antecedent-less pronoun. Indeed, this analysis may account for certain data obtained by McDaniel et al. (1987) who did present sentences such as (17). But it leaves without explanation the fact that in cases in which there is a legitimate linguistic antecedent for the pronoun (as in (2) and (13) in our experiments and in almost all of the previous research mentioned above), children still bind the pronoun locally, consistently or not.

Finally, Grimshaw and Rosen argue that the grammar of emphatic pronouns is in part responsible for Principle B violations. Sentences like (18) are for them acceptable under the indicated indexing.

(18) a. No, Mary$_i$ only likes **HER**$_i$

    b. Mary$_i$ wants to give it to **HER**$_i$
        Grimshaw and Rosen, examples (19)

If children sometimes treat pronouns as emphatic they may provide responses that will incorrectly be interpreted as violations of Principle B while according to Grimshaw and Rosen, emphatic pronouns are not subject to the Binding Theory. Since in the Romance languages clitic pronouns cannot be emphatic (or contrastive), Grimshaw and Rosen predict that children learning one or another of the Romance languages will "obey Principle B more systematically than children learning English" (p. 27). The fact that 3 to 4 years old French speaking children exhibit a considerable success with respect to sentences with pronouns does not contradict this prediction. However, the presence of binding errors with regard to clitic pronouns that are exempt of emphatic construal, constitutes a problem for the hypothesis that the grammar of emphatic pronouns lowers the children's performance on Principle B. It appears that the factor invoked by Grimshaw and Rosen does not entirely explain the observed acquisition lag in Romance (binding errors are also observed in Spanish, cf. Padilla Rivera 1985), even if it may in part account for acquisition data in languages like English.

Thus, the answer to the first question is that the acquisition lag between anaphors versus pronouns is not a consequence of experimental artifacts.

(b) Does the acquisition lag between anaphors versus pronouns in object position justify the claim by Wexler and Chien (op. cit.) that Principle B of the Binding Theory should be reformulated so that it would deal only with bound pronouns, under the view that "pronouns become bound iff they are referentially dependent on a quantifier"? (Montalbetti and Wexler, op. cit.). Once more the answer is no.

Wexler and Chien's hypothesis predicts on the one hand, that at the age at which children's linguistic behavior conforms to Principle A, it will conform to the reformulated Principle B. On the other hand, this hypothesis predicts that children will perform better with pronouns bound by a quantifier than with non-variable pronouns. As for the second prediction, it follows from the assumption that the reformulated Binding Principle, as part of the genetic endowment, is not learned. With respect to the pragmatic ("or whatever") principle that rules out local binding of pronouns when the pronouns are not variables, the authors suggest two different solutions: (i) the pragmatic principle is learned; (ii) it becomes available by maturation.

As the results of our second experiment indicate, both predictions are disconfirmed: there is no difference in the comprehension of non-variable versus variable pronouns (i.e., Figures 8 and 9), and the acquisition lag

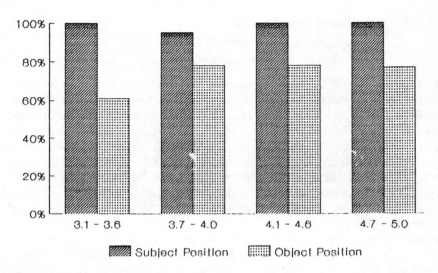

Fig. 8. Percentage of correct responses for non variable pronouns × syntactic position × age.

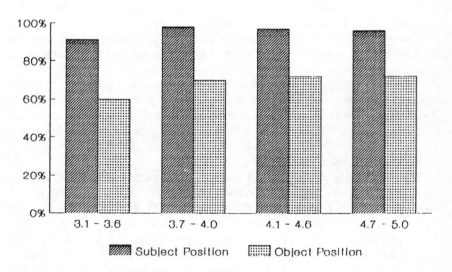

Fig. 9. Percentage of correct responses for variable pronouns × syntactic position × age.

between anaphors versus pronouns in object position persists even if the pronoun is in an environment in which it must be interpreted as a bound variable (i.e., Figure 6). The fact that children perform equally well with variable and non-variable pronouns has also been observed for English by Kaufman (1988). However, Chien and Wexler (1988) found that young

English speaking children reject incorrect interpretations of sentences with a non-variable pronoun less frequently than they reject incorrect interpretations of sentences with a pronoun bound by a quantifier. Since the same children show successful comprehension with regard to both variable and non-variable pronouns when they are asked to verify correct interpretations of the test sentences, the claim that the data first mentioned confirm their prediction must be further qualified (see Grimshaw and Rosen, op. cit.).

Moreover, from a theoretical point of view, it is clear that further research may provide arguments for the reformulation of the Binding Theory as in Chomsky (1981, 1986). However, Montalbetti and Wexler's proposal raises more problems than it solves. As noticed by Grimshaw and Rosen (op. cit.), both the fact that the configurational restrictions for bound variables and non-variables pronouns are exactly the same, and the fact that in the bound variable cases and non-variable cases the target of the constraint is the same, are just accidental. Also, a range of data is not accounted for by Reinhart's pragmatic principle that Chien and Wexler adopt (i.e., cases where there is no bound variable alternative which can be used, cf. Lust 1986). Note finally that embodied in their hypothesis is the idea that "facts and theories drawn from first language acquisition can be crucial in the determination of the grammatical theory" (cf. p. III.7). Nevertheless, until the role of a variety of factors that interact in children's linguistic behavior is better understood, Chien and Wexler's statement is questionable. Acquisition facts per se do not justify a reformulation of the theory of grammar unless they are compatible with well founded arguments concerning the structure of the stable state.

(c) Do the data provide arguments in support of the maturation hypothesis? Once more the answer is no.

If the emergence of UG Principles was dependent on a biologically specified schedule we would expect the same acqusition sequency of these principles across languages and across constructions. However, a comparison of data from different languages reveals that Danish, Dutch and English speaking children perform binding errors with regard to sentences with pronouns until around age 6 or later, while, as seen in this study, French children stop making such errors by the mean age of 3.9 years. This fact indicates that contrary to the maturation hypothesis prediction, cross-linguistic correlations are not attested. Furthermore, data from our second study contradict the cross-construction correlations prediction. If the availability of Principle B was responsible for children's performance with sentences containing a pronoun, we would expect children to master sentences with a pronoun in subject position and sentences with a pronoun in object position at the same age. In the experiment, children's successful comprehension of sentences with either a non-variable or a variable pronoun in subject position provides strong evidence in support of the availability of Binding Principle B. Thus, the fact that within each age group, the children gave correct responses for sentences with a pronoun in subject position significantly more often than for sentences with a pronoun in object position cannot be accounted for by the hypothesis that Principle B is not yet maturationally available (i.e., Figures 8 and 9). The absence of both cross-linguistic and cross-construction correlations constitutes a problem for the maturation hypothesis. Namely, it makes impossible the proposition that the observed acquisition facts can only be due to the unavailability of Principles of Universal Grammar.

Kayne (1988) suggests that in French (Spanish and Italian), an anaphoric clitic binds a VP subject position while the NP that appears in the specifier of IP-position binds the object position.

According to Kayne the representation of (19a) would be as in (19b).

(19) a. Jean se lave
         Jean washes himself

     b. $[_{IP}$ Jean$_i$ se$_j$ $[_{VP}$ t$_j$ [lave t$_i$]]]

To the extent that Kayne's account is correct, constructions with anaphoric clitics, passives and raising are alike in that they involve A-chain formation. The fact that even 3 year olds exhibit successful comprehension of sentences with anaphoric clitics constitutes a problem for the more specific maturational claim according to which A-chain formation matures (cf. Borer and Wexler, op. cit.). (See also Crain, Thornton and Murasagi 1987, Pinker, Lebeaux and Frost 1987, Demuth 1989, Jakubowicz 1989a, for other acquisition data that contradict Borer and Wexler's hypothesis.)

To explain the difference in the amount of correct responses between anaphors and pronouns, I suggested (cf. Jakubowicz 1984) that local A-Binding qualifies as a default principle at a stage when children still do not know which proform falls under which Binding Principles in a given language. This assumption accounts for the behavior of children who uniformly bind anaphors and pronouns, but it fails to explain both, the fact that certain children sometimes consider the pronoun as bound and sometimes as free (attested also in Danish and English), and the fact that in elicited production tasks, Danish children never produce a pronoun when describing a self-oriented action. These two facts lead us to suggest that anaphors are lexically identified first while for certain children, pronouns remain temporarily unidentified and belong to both, the set of anaphors and the set of pronouns (cf. Jakubowicz and Olsen, op. cit.). While the Binding Principles may be ready to apply from the start, they cannot be operative until the child knows how anaphors and pronouns are lexicalized in the language to which he is exposed. The analysis of the production data obtained in this study provides further support for this hypothesis which explains the developmental facts in terms of lexical (and morphological) learning.

As mentioned in section 2.1, in the elicited production task, even the 3 year olds showed a high score of grammatically correct sentences with the reflexive **se** while the 3 to 6 year olds produced a considerable number of grammatically incorrect sentences with a pronoun. Consider again the examples of grammatically incorrect productions with a pronoun, presented in (7) through (10).

It is important to stress that, while sentences with gender errors are structurally wellformed and as such available in the linguistic input, Case and expression-type errors raise a problem. Given that children could not have heard forms such as (7b) and (7c), or (8a) and (8b) in their parent's speech, how can we explain the presence of these forms at the younger ages and their absence at the older?

With respect to the expression-type errors, supplementary data from the same study show that sentences like (7) are not due to an incorrect perception of the scenes represented in the pictures. Further, the hypothesis that young children believe that the clitic **se** is part of the verbal lexical entry is also inadequate (see Jakubowicz 1989a). It appears that very young children know that **se** is a reflexive anaphor. I suggest that at a certain age, the use of **se** instead of **le** ou **la** is due to the fact that the child knows that pronominalization and cliticization are required, but he does not yet know the correct lexical entry to accomplish such processes. The fact that the child moreover names, adds a strong postverbal pronoun, or points to the character recipient of the action indicates that for him, in these cases, **se** alone is insufficient to specify the referential value. A similar explanation can account for the Case errors mentioned above. The child knows that in response to the question he was asked, the R-expressions mentioned in that question can be pronominalized. However, in French, as in English, morphological case marking is exceptional in that it is almost limited to the pronominal paradigm. Given this fact, it is natural to assume that the child will spend some time to learn from experience the ways in which morphological case is realized, and during this time, he will perform errors.

To the extent that the explanation proposed here is on the right track, neither the expression-type errors nor the Case errors, can be said to constitute violations of Principle B or of Case Theory. Rather, these errors reflect the fact that the child needs to figure out how the categorial and agreement features that Universal Grammar makes available are lexicalized in the language to which he is exposed. In this respect note that in French (in Romance more generally), the anaphoric-pronominal distinction is lexicalized only for the third person forms; first and second person forms are ambiguously reflexive or non-reflexive.

Further, third person pronouns constitute a set of overt agreement features in that they are marked for gender, number and case. As it is the case in many languages of the world, when one of these features is combined with others, certain distinctions are neutralized (Greenberg 1966). Thus, in the French pronominal paradigm, gender is neutralized within plural number in the accusative case (**la-le/les**) and within the singular and the plural number in the dative case (**lui/leur**), while there is no gender nesting within number in both the nominative (**elle-il/elles-ils**) and the oblique case (**elle-lui/elles-eux**). Indeed, in determining which notions are encoded morphologically in the pronominal paradigm and which features are neutralized within others, the French child is faced with a considerable search problem. In contrast, given that **se** is invariant for gender, number and case, in other words, since **se** is morphologically underspecified, a similar search problem does not arise in determining the lexical entry for the reflexive anaphor.

The patterns of overgeneralization and free variation among the proforms attested by the production data of this study are equally present in the acquisition of inflexions in a variety of languages (Cf. Slobin 1985). This fact provides independent support for the assumption that the developmental problem raised by the finding that young children are more successful with anaphors than with pronouns in object position can be solved in terms of a lexical (and morphological) learning hypothesis.

To conclude: the data reported here confirm that there is an acquisition lag between anaphors versus pronouns that is restricted to pronouns in object position. As for these last expressions, the fact that children perform better in the sentence-picture matching task than in the act-out task suggests that task related factors must be taken into account in the analysis of the child's responses. In this respect, Grimshaw and Rosen's concerns with regard to the effect of experimental factors is indeed reasonable. However, their proposal that the acquisition lag between anaphors versus pronouns results from experimental artifacts, cannot be accepted. With respect to the maturation hypothesis and the hypothesis according to which Principle B of the Binding Theory should be reformulated, the predictions that follow from these hypothesis have both been disconfirmed. Rather, the computational devices underlying the setting up of anaphoric and disjoint reference relations are already operative when children are about 3 years old. The fact that the performance with pronouns in object position is less perfect than that with anaphors and pronouns in subject position can be explained under the hypothesis that while the Binding Principles are ready to apply from the start, they cannot be correctly implemented until the child has at his command the entire set of anaphors and pronouns of the language to which he is exposed. Results from Danish and from English show that increases in the child's lexicon can equally explain the developmental facts in languages in which anaphors and pronouns are lexicalized differently than in French. It appears then that the invariance of UG Principles plus lexical learning hypothesis of language acquisition should be preferred on theoretical as well as on empirical grounds.

APPENDIX
SENTENCE-PICTURE MATCHING TASK

1. *Test Sentences*

1.1. *Overt Anaphors*

(1) Kiki voit que Nounours se brosse
     Kiki sees that Teddy Bear brushes himself

(2) Nounours dit que Kiki se lave
     Teddy Bear says that Kiki washes himself

(3) Schtroumpfette voit que Barbie se peigne
     Smurf-fem. sees that Barbie combs herself

(4) Barbie dit que Schtroumpfette se mouche
     Barbie says that Smurf-fem. "wipes her nose" (herself)

1.2. *Pronouns*

(5) Schtroumpfette dit que Barbie la brosse
     Smurf-fem. says that Barbie brushes her

(6) Barbie dit que Schtroumpfette la lave
     Barbie says that Smurf-fem. washes her

(7) Nounours dit que Kiki le peigne
     Teddy Bear says that Kiki combs him

(8) Kiki dit que Nounours le mouche
     Kiki says that Teddy Bear "wipes his nose" (him)

2. *Pre-test Sentences*

a. Nounours mange la banane
    Teddy Bear eats the banana

b. Schtroumpfette se mouche
    Smurf-fem. "wipes her nose" (herself)

c. Kiki le peigne
    Kiki combs him

d. Le garçon lave la fille
    The boy washes the girl

e. La fille voit que le garçon est sale
    The girl sees that the boy is dirty

f. La fille écoute le garçon jouer du tambour
    The girl hears the boy playing the drum

g.   Le garçon voit que la fille est triste
     The boy sees that the girl is sad

h.   La fille coiffe le garçon
     The girl combs the boy

ACT-OUT TASK

1. *Test Sentences*

In the sentences below **Jean** and **Claire** represent the name of the subject (a boy or a girl) who performs the experiment. For non variable and variable pronouns in object position, the (a) version is presented only to boys, while the (b) version is presented only to girls.

1.1. *Overt Anaphors*

(1)   Chaperon Rouge veut que Jean (Claire) se peigne
      Little Red Riding Hood wants that Jean combs himself

(2)   Blanche Neige veut que Jean (Claire) se mouche
      Snow White wants that Jean wipes (the nose) himself

(3)   Mickey veut que Jean (Claire) se lave
      Mickey wants that Jean washes himself

(4)   Donald veut que Jean (Claire) se brosse
      Donald wants that Jean brushes himself

1.2. *Non Variable Pronouns in Subject Position*

(5)   Chaperon Rouge dit qu'elle est contente
      Little Red Riding Hood says that she is happy

(6)   Blanche Neige dit qu'elle est endormie
      Snow White says that she is asleep

(7)   Mickey dit qu'il est fatigué
      Mickey says that he is tired

(8)   Donald dit qu'il est triste
      Donald says that he is sad

1.3. *Variable Pronouns in Subject Position*

(9)   Chaque Barbie dit qu'elle est fatiguée
      Every Barbie says that she is tired

(10) Chaque Schtroumpfette dit qu'elle est triste
      Every Smurf-female says that she is sad

(11) Chaque Kiki dit qu'il est endormi
      Every Kiki says that he is asleep

(12) Chaque Nounours dit qu'il est content
      Every Teddy Bear says that he is happy

1.4. *Non Variable Pronouns in Object Position*

(13) a.   Donald veut que Jean le mouche
          Donald wants that Jean wipes him (the nose)

    b.   Blanche Neige veut que Claire la mouche
           Snow White wants that Claire wipes her (the nose)

(14) a.   Mickey veut que Jean le peigne
           Mickey wants that Jean combs him

    b.   Chaperon Rouge veut que Claire la peigne
           Little Red Riding Hood wants that Claire combs her

(15) a.   Donald veut que Jean le brosse
           Donald wants that Jean brushes him

    b.   Blanche Neige veut que Claire la brosse
           Snow White wants that Claire brushes her

(16) a.   Mickey veut que Jean le lave
           Mickey wants that Jean washes him

    b.   Chaperon Rouge veut que Claire la lave
           Little Red Riding Hood wants that Claire washes her

1.5. *Variable Pronouns in Object Position*

(17) a.   Chaque Nounours veut que Jean le peigne
           Every Teddy Bear wants that Jean combs him

    b.   Chaque Barbie veut que Claire la peigne
           Every Barbie wants that Claire combs her

(18) a.   Chaque Kiki veut que Jean le brosse
           Every Kiki wants that Jean brushes him

    b.   Chaque Schtroumpfette veut que Claire la brosse
           Every Smurf-female wants that Claire brushes her

(19) a.   Chaque Nounours veut que Jean le lave
           Every Teddy Bear wants that Jean washes him

    b.   Chaque Barbie veut que Claire la lave
           Every Barbie wants that Claire washes her

(20) a.   Chaque Kiki veut que Jean le mouche
           Every Kiki wants that Jean wipes him (the nose)

    b.   Chaque Schtroumpfette veut que Claire la mouche
           Every Smurf-fem. wants that Claire wipes her (the nose)

2. *Fillers*

(1) Donald veut que Jean (Claire) caresse le chat
     Donald wants that Jean caresses the cat

(2) Mickey veut que Jean (Claire) pousse la voiture
     Mickey wants that Jean pushes the car

(3) Blanche Neige veut que Jean (Claire) boive le coca
     Snow White wants that Jean drinks the coke

(4) Chaperon Rouge veut que Jean (Claire) cache la balle
     Little Red Riding Hood wants that Jean hides the ball

(5) Chaque Nounours veut que Jean (Claire) touche Donald
     Every Teddy Bear wants that Jean touches Donald

(6) Chaque Kiki veut que Jean (Claire) frotte Mickey
Every Kiki wants that Jean rubs Mickey

(7) Chaque Barbie veut que Jean (Claire) essuie Chaperon R.
Every Barbie wants that Jean dries Little Red R. Hood

(8) Chaque Schtroumpfette veut que Jean (Claire) gratte Blanche Neige
Every Smurf-fem. wnats that Jean scratches Snow White

3. *Pretest Sentences*

a. Chaque Nounours dance
Every Teddy Bear dances

b. Chaque Kiki fait des galipettes
Every Kiki does a somersault

c. Chaque Schtroumpfette fait des galipettes
Every Smurf-female does a somersault

d. Chaque Barbie saute
Every barbie jumps up and down

e. Chaque Nounours est content
Every Teddy Bear is happy

f. Chaque Barbie est fatiguée
Every Barbie is tired

g. Chaque Kiki est triste
Every Kiki is sad

h. Chaque Schtroumpfette est endormie
Every Smurf-fem. is asleep

i. Chaque Kiki peigne Chaperon Rouge
Every Kiki combs Little Red Riding Hood

j. Chaque Barbie essuie Donald
Every Barbie dries Donald

k. Blanche Neige lave chaque Nounours
Snow White washes every Teddy Bear

l. Mickey mouche chaque Schtroumpfette
Mickey wipes every Smurf-female

REFERENCES

Borer, Hagit and Kenneth Wexler: 1987, 'The Maturation of Syntax', in T. Roeper and E. Williams (eds.), *Parameter Setting*, Reidel, Dordrecht, 123—172.

Borer, Hagit and Kenneth Wexler: 1988, 'The Maturation of Grammatical Principles', paper presented at the GLOW Colloquium, Budapest.

Chien, Yu-Chin and Kenneth Wexler: 1987, 'Children's Acquisition of the Locality Conditions for Reflexives and Pronouns', *Papers and Reports on Child Language Development 26*, Stanford University.

Chien, Yu-Chin and Kenneth Wexler: 1988, 'Children's Acquisition of Binding Principles', paper presented at the Thirteen Annual Boston University Conference on Language Development.

Chomsky, Noam: 1965, *Aspects of the Theory of Syntax*, MIT Press, Cambridge, Massachusetts.

Chomsky, Noam: 1981, *Lectures on Government and Binding*, Foris, Dordrecht.

Chomsky, Noam: 1986, *Knowledge of Language: Its Nature, Origins and Use*, Series Convergence, Praeger, New York.

Clahsen, Harald: 1988, 'Learnability Theory and Problem of Development in Language Acquisition', manuscript, Allgemeine Sprachwissenschaft Universität, Düsseldorf.

Crain, Stephen and Janet Fodor: 1987, 'Competence and Performance in Child Language', paper presented at the Tel Aviv Fifth Annual Workshop in Human Development and Education.

Crain, Stephen, N. Thornton and K. Murasagi: 1987, 'Capturing the Evasive Passive', paper presented at the 12th Annual Boston University Conference on Language Development.

Demuth, Katharine: 1989, 'Maturation and the Acquisition of the Seshoto Passive', *Language* **65**(1), 56—81.

Deutsch, Werner, Charlotte Koster and Jan Koster: 1986, 'What Can We Learn from Children's Errors in Understanding Anaphora?', *Linguistics* **24**, 203—225.

Felix, Sascha: 1984, 'Maturational Aspects of Universal Grammar', in A. Davis, C. Criper and A. Howat (eds.), *Interlanguage*, Edinburgh University Press, Edinburgh.

Felix, Sascha: 1988, 'Universal Grammar in Language Acquisition', manuscript, University of Passau.

Greenberg, Joseph: 1966, *Universals of Language*, MIT Press, Cambridge, Massachusetts.

Grimshaw, Jane and Sarah Rosen: 1990, 'Knowledge and Obedience: The Developmental Status of the Binding Theory', *Linguistic Inquiry* **2**(2), 187—223.

Jakubowicz, Celia: 1984, 'On Markedness and Binding Principles', *Proceedings of the North Eastern Linguistic Society, 14*, pp. 154—182.

Jakubowicz, Celia: 1987, 'Contraintes syntaxiques et acquisition du langage', *Annales de la Fondation Fyssen* **3**, 69—83.

Jakubowicz, Celia: 1989a, 'Maturation or Invariance of Universal Grammar Principles in Language Acquisition', in A. Giorgi and B. Dotson Smith (eds.), *Romance Languages and Linguistics, Probus* **1—3**, 283—340.

Jakubowicz, Celia: 1989b, 'Invariance of Universal Grammar Principles in the Acquisition of Reflexives, Anaphors, Passive, Promis and Raising Constructions in French', paper presented at the Fourteen Annual Boston University Conference on Language Development.

Jakubowicz, Celia and L. Olsen: 1988, 'Reflexive Anaphores and Pronouns in Danish: Syntax and Acquisition', paper presented at the Thirteen Annual Boston University Conference on Language Development.

Kaufman, Diana: 1988, *Grammatical and Cognitive Interactions in the Study of Children's Knowledge of Binding Theory and Reference Relations*, Ph.d. dissertation, The Temple University.

Kayne, Richard: 1988, 'Romance SE/SI', paper presented at the GLOW Colloquium, Budapest.

Lust, Barbara (ed.): 1986, *Studies in the Acquisition of Anaphora: Defining the Constraints*, Reidel, Dordrecht.

McDaniel, Dana, H. Smith Cairns and J. Ryan Hsu: 1987, 'Binding Principles and Control in Children's Grammars', manuscript, CUNY and The William Paterson College of New Jersey.

Montalbetti, Mario and Kenneth Wexler: 1985, 'Binding Is Linking', *Proceedings of the West Coast Conference on Formal Linguistics, 4*, pp. 228—245.

Otsu, Yukio: 1981, *Universal Grammar and Syntactic Development in Children: Toward a Theory of Syntactic Development*, Ph.d. dissertation, MIT.

Padilla-Rivera, Jose: 1985, *On the Definition of Binding Domains in the First Language Acquisition of Anaphora in Spanish*, Ph.d. dissertation, Cornell University, Ithaca, New York.

Pinker, Steven: 1984, *Language Learnability and Language Development*, Harvard University Press, Cambridge, Massachusetts.

Pinker, Steven, David S. Lebeaux and L. Frost: 1987, 'Productivity and Constraints in the Acquisition of the Passive', *Cognition* **26**, 195—267.

Reinhart, Tanya: 1983, 'Coreference and Bound Anaphora: A Restatement of the Anaphora Questions', *Linguistics and Philosophy* **6**, 47—88.

Slobin, Dan: 1985, 'Crosslinguistic Evidence for the Language-making Capacity', in D. Slobin (ed.), *The Crosslinguistic Study of Language Acquisition*, Vol. 2, Erlbaum, Hillsdale, New York.

Solan, Larry: 1987, 'Parameter Setting and the Development of Pronouns and Reflexives', in T. Roeper and E. Williams (eds.), *Parameter Setting*, Reidel, Dordrecht, 189—210.

Wexler, Kenneth and Yu-Chin Chien: 1985, 'The Development of Lexical Anaphors and Pronouns', in *Papers and Reports on Child Language Development*, Stanford University.

SHYAM KAPUR, BARBARA LUST, WAYNE HARBERT
AND GITA MARTOHARDJONO

UNIVERSAL GRAMMAR AND LEARNABILITY THEORY: THE CASE OF BINDING DOMAINS AND THE 'SUBSET PRINCIPLE'[1]

1. INTRODUCTION

Recently it has been argued that the acquisition of language knowledge depends critically on one principle of a general cognitive 'learnability module', the 'Subset Principle' (SP). This principle, which derives from studies of computational induction in a framework of recursive function theory (Angluin 1980), has now been claimed to be a 'necessary condition for (knowledge) acquisition from positive only evidence' (Berwick 1982, 1985). In this paper, we critically investigate the role of this learning principle in the acquisition of language knowledge. In particular, we will consider the recent argument that an informal version of the SP constrains both first language acquisition of anaphora and the formal linguistic theory of anaphora, i.e., Binding Theory (BT) itself. (Berwick 1982, 1985, Jakubowicz 1984, Manzini and Wexler 1987.)

In another paper (Kapur et al. 1990), we consider the question of the connection between this informal SP and its counterpart in formal learning theory, and conclude that it is neither equivalent to nor entailed by the original formal learning theory from which it is descended. We consider the relation of this proposal to a strong form of a theory of UG which does not consult a learnability module. In the present paper, however, we will restrict ourselves to developing certain arguments which bear on the status of the informal SP as an empirical hypothesis about the representation and acquisition of natural language. We will be particularly concerned with its predictions with regard to BT (which determines the distribution and interpretation of pronouns and anaphors).

We have chosen to focus on the form of the SP developed in Manzini and Wexler 1987, since the hypothesis finds its most precise and detailed statement, as well as its most extensive application to natural language data, in that work. In particular, we will focus on Manzini's and Wexler's (M & W) most interesting application of the SP which involves the proposal that Governing Categories (GC) of the sort that play a role in the Binding Theory are parameterized in accord with the SP in linguistic theory, and that their acquisition is constrained by this principle. M & W propose that cross-linguistic variation in GC for anaphors and pronouns can be represented in terms of an inclusion hierarchy, and that the child's

185

Eric Reuland and Werner Abraham (eds), Knowledge and Language, Volume I, From Orwell's Problem to Plato's Problem: 185—216.

knowledge of a local Binding Domain on this hierarchy is induced from computations of language size, which consult the SP.

The central empirical prediction of this proposal is that opposite hierarchies of markedness result from the SP for pronouns and anaphors. This is where the SP proposal can be distinguished crucially from a UG theory of markedness. As M & W recognize: 'An especially impressive demonstration of modularity would exist if the learning module predicted different markedness hierarchies for the same parameter depending upon cases, for a substantive assumption of markedness within linguistic theory would not do this' (1987, p. 41); 'To the extent that empirical evidence corroborates the different hierarchies, this provides evidence for the relevance of the Subset Principle' (1987, p. 46). In the present study, we will focus on this prediction. In the first part of our paper we will adduce evidence from cross-linguistic variation in Binding Domains which disconfirms this proposal empirically. We will show that an attempt to deal with such data by hypothesizing an additional stipulation on the SP proposal, viz., the 'Spanning Hypothesis', fails. This evidence suggests that the SP does not sufficiently restrict the linguistic theory of local Binding Domains; rather a strong theory of UG does. In the second part of the paper we will demonstrate that the results of research on first language acquisition are consistent with our cross-linguistic findings; but not with the SP.

2. THE SUBSET PRINCIPLE AND BINDING THEORY

In their recent extension of the informal SP to the study of the BT, M & W (1987) propose that cross-linguistic variation in Binding domains can be described by a five value parameter, whose values stand in a subset/superset relation with respect to each other. This hierarchy is given in (1):

(1)   (= M & W 29)
        γ is a governing category for α iff γ is the minimal category that contains α and a governor for α and

   a. can have a subject, or, for α = anaphor, has a subject β, $\beta \neq \alpha$; or
   b. has an INFL; or
   c. has a Tense; or
   d. has a 'referential' Tense; or
   e. has a 'root' Tense

(if, for α anaphoric, the subject β' ($\beta' \neq \alpha$) of γ, and of every category dominating α and not γ, is accessible to α.)

English is argued to provide an example of GC(a) for an anaphor (e.g. 'himself'); Japanese is argued to provide an example of GC(e) for an anaphor (e.g. 'zibun'). A second parameter, the Proper Antecedent Parameter, determines whether all c-commanding arguments or only subject arguments count as possible antecedents for anaphors and pronouns.

M & W (1987) are concerned with determining how learners cope with the 'subset problem' — what prevents a learner of English, for example, from guessing value GC(e) on this hierarchy as the correct value for anaphors — a guess which will result in overgeneration, and from which s/he could retreat only on the basis of presumably unavailable negative evidence. Their answer is that overgeneration is prevented in this case because learning is guided by the Subset Principle (2) (1987, p. 425).

(2)   Subset Principle: Let p be a parameter with values $p_1, \ldots, p_n$, f_p a learning function and D a set of data. Then for every p_i, $1 \leq i \leq n, f_p(D) = p_i$ if and only if

    a. $D \subseteq L(p_i)$ and
    b. for every $p_j, 1 \leq j \leq n$, if $D \subseteq L(p_j)$, then $L(p_i) \subseteq L(p_j)$

M & W interpret this SP as requiring that '... given two languages, one of which is a subset of the other, if both are compatible with the input data, ... the learning function must pick the smaller one' (p. 414). M & W assume two other conditions which are crucial if learning guided by the SP is to succeed. The first of these is the Subset Condition (SC), which requires that the values of the parameter in question generate languages which are in a subset relation with respect to each other (1987, p. 429). The second of these is the Independence Principle, which requires that 'the subset relations that are determined by the values of a parameter hold no matter what the values of the other parameters are,' i.e., that the choice of value for one parameter may not depend on the choice of value for a second one (1987, p. 434).

M & W's main empirical argument for the proposal that the subset principle plays a role in the determination of parameter values involves the claim that cross-linguistic variation in the formulation of governing category can be extensionally characterized in terms of subset relations, and therefore satisfies the Subset Condition. Since the SC is viewed as a precondition for the Subset Principle to 'determine' learning (1987, p. 429), this state of affairs is explained if the cross-linguistic parameterization of GC's is in fact determined by the subset principle.[2]

Parameter values such as those in (1) are set for individual lexical items, under the Lexical Parameterization Hypothesis of M & W. They propose that the definition of GC for anaphors and pronouns can vary as described in (1) both within and across languages, for individual lexical items. The subset relations among the values can be translated into markedness relations, with different results for anaphors than for pronouns. In this theory, the extensionally smallest language is the least marked one. Value (a) generates the smallest subset language for anaphors, while (e) generates the smallest for pronouns. Thus, while GC(a) is the least marked value for anaphors, GC(e) is the least marked for pronouns.

The logic here can be illustrated by reference to (3).

(3) $[_D \ldots a2 \ldots [_C \ldots a1 \ldots b \ldots]]$

In a language X where anaphors are bound in domain D, both antecedents a1 and a2 can bind an anaphor 'b'. In a language Y where anaphors must be bound in the domain C, b may only be bound to a1. Language Y is thus extensionally smaller than X; i.e., it has fewer sentences (indexed strings) in it. However, in a language X where pronouns must be free in domain D, b may not corefer with either a1 or a2. In a language Y where pronouns need to be free only in the domain C, b may corefer with a2. In this case, language Y is extensionally larger than language X, containing one more indexed string than X. Figure 1 displays these reverse subset relations.

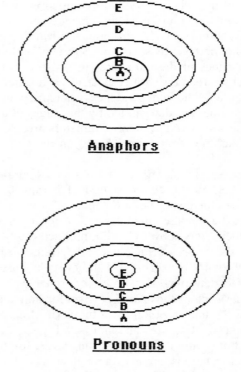

Fig. 1. Hierarchy of binding domains.

It is clear that this markedness proposal for GC's for anaphors and pronouns is distinguished from that in a strong theory of UG (e.g. Chomsky 1981a, b), in which the 'near complementary' distribution of anaphors and pronouns is derived as a consequence of their shared and syntactically 'minimal' domain. In most treatments, this syntactically minimal domain would correspond to what M & W have described as GC(a). There have been attempts within the theory of UG to account for apparent discrepancies to the complete complementarity of anaphors and pronouns. However, as in Huang 1983, for example, although such attempts propose to differentiate the domains of anaphors and pronouns, they typically maintain syntactic minimality for both (cf. Chomsky 1986). While the SP proposal is clearly an extensional proposal which refers to computation of language size, the UG proposal is clearly an intensional proposal referring to properties of grammar in order to determine markedness (Chomsky 1986). (See Kapur et al. (1990) for contrast of intensional and extensional characterization of markedness. See Gair (1988) for study of intensional UG-based approaches to markedness; and Koster 1981.)

In the SP proposal, similar reasoning leads to the same mirror-image situation with respect to a second proposed parameter, the antecedent parameter, given in (4):

(4)   (= M & W 50)
        A proper antecedent for α is

        a. any subject β; or
        b. any element β.

For anaphors, value (a) is the unmarked one, while for pronouns, value (b) is the unmarked one, since these are again the values yielding the smallest sets of indexed strings.

A further restriction, given in (5), is stipulated to prohibit pronouns and anaphors from assuming the marked value for both parameter (1) and parameter (4).

(5)   (= M & W 56)
        An anaphor must be bound in either its unmarked governing category or by its unmarked proper antecedent.
        A pronoun must be free either in its unmarked governing category or from its unmarked proper antecedents.

The latter principle is included to capture the fact that (in some languages) 'long distance' reflexives allow only subject antecedents, even when, as is the case in Icelandic, reflexives bound to clause-mate antecedents allow both subject and object antecedents for some speakers.[3] The existence of (5) is claimed to avoid the need for explicitly linking the choice of value for binding domain to the choice of value for proper antecedent.[4] Such a linking would violate the Independence Principle which is proposed to be a precondition for the operation of the Subset Principle.

In this paper, we will assume the above set of conditions on the SP proposal for the purpose of argument, and confront the critical empirical prediction of this proposal. We note, however, that aside from the questionable status of (5) as a principle of UG, it appears that it cannot be maintained on empirical grounds, since there do seem to be languages, such as Korean, in which long-distance (LD) anaphors may be object bound (Lee 1988). Thus the problem of accounting for the restriction in (5) remains in languages where it does hold.

3. AN EMPIRICAL PROBLEM FOR THE SUBSET PRINCIPLE

A major empirical problem with this extensional approach to markedness in the fixing of binding domains is posed by the fact that, as M & W note (1987, p. 439), pronouns tend across languages to be associated with the 'marked' values of the parameter in (1), not with the unmarked values, where markedness is determined by the SP. This fact critically disconfirms the essential empirical prediction of the SP. In particular, pronouns tend to be required to be free only in what is characterized as domain (a) in (1). As reported in Harbert (1986a, b) and elsewhere, this value is quite frequent across languages. It is the one observed for pronouns in New Testament Greek, Khmer, Swahili, Dogrib, Japanese, Colloquial Russian, Chinese, Imbabura Quechua, Basque and Malayalam, for example. In all of these languages, pronouns must be free in the domain of the minimal subject, but may be bound outside that domain, as in (6)—(15). See Harbert (1986b) for further development of this claim.

(6)   New Testament Greek (Luk 8:37)
       Kaì ērōtēsen autòn$_j$ *hapan to plēthos* . . .   [$_S$ PRO$_j$
       and asked   him   all   the population

       apetheîn ap' *autōn/(*heautōn)*]
       go   from them$_i$/selves$_i$

(7)   Khmer (Fisher 1984)
       Puu-Sok   thaa [$_S$ qom-Rɨn   khəəñ *koət/*kluən*]
       Uncle Sok$_i$ say   Uncle Rin see   him$_i$ *self$_i$

(8)   Swahili (I. Alemán, personal communication)
       Hamisi alimwamrisha Asadi$_j$ [$_S$ PRO$_j$kumnyaa {*0/yeye*}]
       Hamisi$_i$ ordered   Asadi   INF-shave him$_i$

(9)   Japanese (Oshima 1979)
       John-wa   Mary$_j$-ni   [$_S$ PRO$_j$ *kare*-ni
       John$_i$-TOP Mary$_j$-DAT   him$_i$-DAT

       denwa-o kake]-sase-ta
       to-telephone   caused

(10)   Colloquial Russian (Yokoyama 1980)
       Tanju$_j$ [$_S$ PRO$_j$ nalit'   *ej*   vody] poprosila
       Tanja-ACC   to-pour her$_i$ water asked

       mat'
       mother$_i$-NOM

(11)   Chinese (Huang 1982)
       Zhangsan shuo [$_S$ ni   kanjian-le *ta*]
       Zhangsan$_i$ say   you   see-ASP   him$_i$

(12)   Imbabura Quechua (P. Cole, personal communication)
       Jose crin   [$_S$ Maria *pay*-ta   rikushka-ta]
       José$_i$ believes   Maria him$_i$-Acc see-Nominal-Acc

(13)   Dogrib (L. Saxon 1983)
       John sìi   [$_S$ we-gha ʔelà whihtsi] yek'èrèzho
       John$_i$ FOC   him$_i$-for boat I-made   he-knows

       "John knows that I made a boat for him."

(14)   Basque (Echavarri-Dailey 1986)
       Jonek [Mirenek *bera/*bere burua* iltzea] ez du nahi
       John$_i$ Mary   him$_i$/*self$_i$ to-kill not want

(15)   Malayalam (Mohanan 1982)
       moohan paraññu [meeři *awane* aařaadhik'k'uɲɲu]
       Mohan$_i$ said   Mary him$_i$ worshipped

On the other hand, GC(e), the putative least marked value for pronouns, according to the SP, seems to be rarely attested, if at all. Manzini and Wexler introduce two possible cases. One of these is the Yoruba subject pronominal clitic, for which they refer to the discussion in Mohanan (1982). The more recent discussion of this form in Pulleyblank (1985), however, suggests that it is not a pronoun, but a clitic associated with a null-operator-bound variable, whose disjoint reference properties follow from Principle C, not Principle B.[5] We believe similarly that the second alleged instance adduced by Manzini and Wexler of pronouns which must be free in all domains — that of English anaphoric epithets such as 'the idiot' — are better treated as names, whose referential properties derive from Principle C.[6] In any case, pronouns observing GC(e) are· at least infrequent if they exist at all, and we know of none observing GC(d) or GC(b). Manzini and Wexler cite Icelandic as a language instantiating GC(c) for pronouns — a claim with which we are in agreement. Such languages, however, also seem to be quite rare.[7]

4. THE SPANNING HYPOTHESIS

In part to account for the frequency of occurrence of the marked value GC(a) for pronouns, M & W introduce the Spanning Hypothesis (SH) in (16):

(16)   Spanning Hypothesis: Any given grammar contains at least an anaphor and a pronominal that have complementary or over-lapping distribution.

This hypothesis predicts that 'if [an antecedent] α c-commands β, then α can bind β, where β is an anaphor or a pronominal, or both.' That is, there should be no 'binding gaps' in any language — no constructions in which neither a pronoun nor an anaphor in a given position can be bound to a c-commanding antecedent. This, they claim, explains why pronouns are frequently associated with marked values of the parameters in question. Given the Spanning Hypothesis in (16), the GC for a pronoun can never be larger than that for an anaphor, assuming a language with only one of each, since under those conditions there would be illicit 'un-spanned' domains in which neither could occur. Accordingly, if the reflexive in a language has the least marked domain, according to the SP, i.e., GC(a), the pronoun is forced to a marked value for pronouns according to the SP, i.e., GC(a).

4.1. *Evaluation of the Spanning Hypothesis*

The SH suffers from both logical and empirical problems.

4.1.1. *Logical Problems*

Fundamentally, the SH vitiates the critical empirical prediction of the hypothesized learnability module by overriding the predicted opposite unmarked values for anaphors and pronouns. It forces either 'anaphor' or 'pronoun' to be marked in terms of the SP, if one is unmarked. As we saw above, it is precisely these empirical predictions that would be necessary to distinguish the SP proposal from a theory of markedness in UG. Logically, it also appears to be incoherent with M and W's 'Independence Principle' and their 'Lexical Parameterization Hypothesis', since it has the effect of creating a partial dependency between choices for domains and choices for antecedents; and it links two lexical items.

Furthermore, note that the SH does not itself specify whether the 'anaphor' or 'pronoun' domain is set first and there is nothing in the general theory of the SP which would do so. If the anaphor domain is set first, then the SH predicts that both anaphor and pronoun should take GC(a). Note that this has the consequence that the SP is no longer the sole determinant of learning in the case of pronouns. Where GC(a) is chosen for the anaphor(s) in a language, pronouns can have no other value. It also has the consequence that the SP now makes the same predictions as a strong theory of UG, i.e. the GC for anaphor and pronoun is the same, and it is syntactically minimal.

On the other hand, if the pronoun domain is established first, given the SH, then we should expect that it should tend to take the putative 'unmarked' value according to the SP, i.e., GC(e), forcing the anaphor domain to adopt the same value. Crucially, we should therefore expect there to be languages in which pronouns must be free in the GC(e) domain, and which also exhibit long distance anaphora. Such a situation might be expected, in fact, to be at least as frequent as one in which both pronouns and anaphors observe value GC(a). Our investigation indicates that this is clearly not the case, as seen in section 3 above. Even languages ostensibly obeying GC(e) for anaphors, e.g. Japanese, typically obey GC(a) for pronouns.

4.1.2. *Empirical Problems*

In addition, the SH is empirically insufficient even when interpreted as though the anaphor value must be set first. Manzini and Wexler claim that their GC parameter, in conjunction with the SH, entails that 'at least one pronominal in a language cannot be associated with the unmarked values of both binding theory parameters . . .', i.e., the GC parameter (in 1 above) and the Proper Antecedent (PA) parameter (in 4 above). This result does seem to follow if Principle (5) is also assumed. Suppose that the pronoun in some language did assume the unmarked value for both parameters — that is, that it had to be disjoint in reference with both subjects and objects in the domain of GC(e). In such a case, the Spanning Hypothesis could be satisfied only if that language also contained an anaphor which could be bound to both subjects and objects in GC(e), preventing binding gaps. In such a case, however, the anaphor would be marked with respect to both parameters — a violation of (5), if UG abides by such a principle.

Consider, however, another case, in which the (marked) value (a) of PA parameter is chosen for the pronoun in some language — i.e., in which pronouns must be disjoint in reference only with respect to subjects. Suppose further that the value chosen for GC for the reflexive in that language is a marked one, e.g. GC(e). In this case, we would expect the pronoun to be free to choose the least marked value for the GC parameter, GC(e), since no violation of the Spanning Hypothesis nor of the proposed Principle (5) would arise. Thus, we are led to expect that languages with long distance reflexive binding of anaphors should tend to have long distance disjoint reference of pronouns as well, at least as often as not.

To see the logic of this, consider more closely the case where the reflexive in a language has chosen value GC(e) for the GC parameter and value (b) for the PA parameter. (It follows from (5) that the choice of value for the PA parameter will be identifiable only in instances of binding within GC(a). In instances of binding a reflexive within larger (more marked) domains, (5) insures that only subject antecedents will be allowed.) In such a case, the pronoun would be able to observe GC(e) too if it also observed the marked value (a) of the PA parameter. Choice (e) for the GC parameter and (b) for the PA parameter would yield binding gaps, since higher clause objects could neither bind reflexives in lower domains (because of (5)) nor corefer with pronouns in those domains. Value (b) of the PA parameter could be chosen for the pronoun just in case GC(a) were also chosen. The following distribution would result, which satisfies the Spanning Hypothesis:

(17) NP_i NP_j $[NP_k$ NP_l Pron $i/j/{*}k/{*}l$]
                                                REFL $i/{*}j/k/l$

        subject object subject object

In such a situation, the pronoun would be forced to be marked either with respect to the choice of GC or with respect to the choice of antecedent. Nothing in M & W's account provides any reason for preferring one of these resolutions over the other.

Consider next the case where the reflexive chooses GC(e) and value (a) for the PA parameter. The latter choice is compatible only with a corresponding choice of value (a) for the PA parameter for pronouns, since otherwise binding gaps would arise in simple sentences. In a structure like (18), the object NP, NP_i, could not bind α whether α was a pronoun or an anaphor.

(18) $[NP .. NP_i ... \alpha]$

Thus, we may consider the choice of value for the PA parameter for the pronoun in this case to be forced by the choice for the anaphor, given the SH. In such a case, the pronoun could then in principle choose either GC(a) or GC(e) for parameter (1). Under the latter choice, pronouns would have to be free from subjects in higher domains, but no binding gaps would arise since anaphors can be bound to subjects in higher domains. In this case again then, since GC(e) is the unmarked choice for pronouns, we should expect it to be the more frequent choice, even given the SH.

Surprisingly, these expectations are not fulfilled. In such LD reflexive languages as Chinese, Japanese, Malayalam and Korean, where reflexives seem to observe GC(e), and where pronouns could therefore choose the least marked value for GC without violating the Spanning Hypothesis, they do not do so, but rather continue to be associated with what is, according to M & W, the most marked value — GC(a). That is, pronouns tend to be associated with GC(a) even where apparently not forced to do so by independent considerations arising from the SP supplemented by the SH. This suggests that what corresponds to GC(a), the syntactically minimal domain, is indeed the unmarked value for pronouns as well as anaphors.

4.1.3. *The SP and Syntactic Change*

The contention that GC(a) is the unmarked domain for pronouns, as well as anaphors, contrary to the prediction of the Subset Principle approach, is also motivated by certain diachronic developments. Harbert (1983, 1986a) argues that in a number of independent cases in the Germanic languages, the GC value for pronouns has changed historically from GC(c) to GC(a), typically, (but not always), paralleled by corresponding changes in anaphoric domains. We illustrate these developments here by contrasting Gothic in (19a, b, c), representative of older Germanic, with German in (19d, e). See Harbert (1986a) and the appendix to the present paper for detailed discussion and examples from other languages.

(19) a. *þai*-ei ni   wildedun [S mik þiudanon ufar
       who$_i$ not they-wanted me to-rule   over

   sis/(*im)] (Luk 19:27)
   selves$_i$/*them$_i$

   "who didn't want me to rule over them"
   (Greek has pron)

   b. jah bedun ina$_j$ *allai   gaujans* . . .
      and asked him all of the inhabitants$_i$

   [S PRO$_j$ galeiþan fairra *sis*] (Luk 8:37)
          to go from selves$_i$

   "And all of the inhabitants asked him to go from them."
   (Greek has pron)

c. (*is*) qaþ-uh   -þan [NP þamma$_j$ [S PRO$_j$
   he$_i$ said-and then   to-the-one

   haitandin *sik*/*ina*] (Luk 14:12)
   inviting self$_i$/*him$_i$

   "And then he said to the one inviting him . . ."
   (Greek has pron)

d. *Sie* baten ihn$_j$ [PRO$_j$ von *ihnen*/**sich*   zu gehen]
   They$_i$ asked him   from them$_i$/*selves$_i$ to go

e. *Er* sagte dem   [PRO *ihn*/**sich* einladenden] Mann . . .
   He$_i$ said to-the   him$_i$/*self$_i$ inviting   man

If we assume that GC(a) is the unmarked domain for pronouns, then the
occurrence of this kind of change might be accounted for, under a
plausible extension of markedness theory, as a change from a marked
grammar to a less marked, and therefore more highly valued one. On the
other hand, given an extensional definition of markedness, the change in
pronominal GC's observed here would be a change in the direction of
greater markedness, i.e. larger subsets, and therefore not directly explain-
able.

4.1.3.1. *The SH and Syntactic Change.* On first consideration, it seems as
if we might imagine that in these languages the change began with
reflexives, which changed from GC(c) to GC(a), i.e., in the direction of
diminished markedness, and that the parallel change in GC for pronouns,
from larger and less marked to smaller and more marked, was driven by
the need to avoid violation of the Spanning Hypothesis. In fact, however,
there is some evidence that this is at least not always the order in which
such events take place. Comparative evidence suggests that we can reconstruct for Norwegian an earlier stage at which GC(c) was the GC for
pronouns, as it was in Gothic and, apparently, Old Norse, and as it still is
in the closely related Icelandic. However, as the example in (20) shows,
pronouns are now required to be free only in GC(a), that is, the domain of
a subject. Crucially, this change occurred without a corresponding change
in the GC for simple reflexives, which, as (21) shows, are still associated
with the value GC(c). In this instance, the shrinking of the GC for
pronouns has clearly not been forced by the Spanning Hypothesis. It is
likely that a similar argument could be made for Russian, where a similar
asymmetry in domains exists between pronouns and reflexives, as shown
in (21), though our acquaintance with the history of these constructions is
insufficient to allow us to make confident claims here.

(20)   *Ola* bad oss [PRO snakke om *seg/ham*]   (Bresnan 1983)
        Ola$_i$ asked us   to-talk about self$_i$/him$_i$

(21)   *mat'* poprosila doč   [PRO nalit'
       mother$_i$ asked daughter   to-pour

       sebe/ej vodu] (Faltz 1977, p. 18)
       self$_i$/her$_i$ water

A similar argument can be constructed on the basis of facts from Old English. Comparative evidence allows us to reconstruct a stage in pre-Old English in which pronouns occurring as possessive specifiers in NP's could not corefer with the subjects of the clauses containing those NP's — i.e., a stage at which NP's did not function as GC's for pronouns in those positions. This was again the case in Gothic (cf. appendix, ex. 2), and is still the case in Modern Icelandic, as the example in (22a) shows. By the time of attested Old English, however, NP had come to function as a binding domain for pronouns, so that they could corefer with the matrix subject. This is seen in (22c). That is, the domain for pronouns has again gotten syntactically smaller, but extensionally larger, contrary to the expectations derivable from markedness according to the SP. Moreover, this development again cannot be accounted for by the Spanning Hypothesis, since at the same time the reflexive was still allowed to occur in this context, as shown in (22b).

(22) a. *Jón* rétti Haraldi [$_{NP}$ *sín/*hans* föt] (Thráinsson 1976)
         John$_i$ handed Harold self's$_i$/*his$_i$ clothes

     b. and him *Hróðgar* gewāt to [$_{NP}$ hofe *sīnum*] (Bēowulf 1236)
         and Hrothgar$_i$ betook himself to his$_i$ house.

     c. (*he*) sealde [$_{NP}$ *his* hyrsted sweord]
         he$_i$ gave   his$_i$ decorated sword

         . . . ombihtþegne (Bēowulf 672)
             to-servant

An essentially identical development in certain varieties of Danish is discussed in Harbert (1983).

5. LINGUISTIC CONCLUSIONS: INTENSIONAL MARKEDNESS

These results in which GC(a) is picked out as the unmarked binding domain for both anaphors and pronouns in synchronic as well as diachronic evidence would be predicted under an account in which markedness is determined intensionally, in terms of minimal syntactic domain, rather than extensionally, in terms of language size.

6. THE SUBSET PRINCIPLE AND ACQUISITION

The opposite hierarchies determined by the SP for anaphors and pronouns are also intended to impose empirical consequences for the course of first language acquisition of anaphora (cf. Berwick 1982, 1985). In this section we review selected empirical studies which suggest that similar to the synchronic and diachronic linguistic data we have seen above, evidence from acquisition supports an intensional UG theory of markedness, rather than an extensional theory of markedness. The acquisition data also provide evidence for the same syntactically minimal binding domain for both anaphors and pronouns in children's early hypotheses about the language they are learning.

Elsewhere, we consider several problems with the general empirical predictiveness of the SP for study of first language acquisition (Kapur et al. 1990, Lust et al. 1989, in preparation).[8] However, for the purposes of this paper we will consider one interpretation of the SP proposal which appears to make precise predictions regarding markedness. Under this interpretation, the SP predicts the child's initial hypothesis for the binding domain of a lexical item which is an anaphor, to be distinct from that for the binding domain of a lexical item which is a pronoun. In particular, as we have seen above, the SP defines GC(a) as the most unmarked hypothesis for anaphors, since it corresponds to the extensionally smallest set in (1). As we have suggested, however, this GC also corresponds to the syntactically most restrictive domain and hence would also constitute the initial hypothesis in a UG theory of markedness defined syntactically. In its predictions for the binding domains of anaphors then, the SP proposal seems to be indistinguishable from a UG theory of markedness. Both predict the most restrictive syntactic domain to be the child's initial hypothesis for BT.

Consider however, the case of pronouns: here the SP predicts the child's initial hypothesis to be GC(e), as seen in Figure 1. Since this domain corresponds to the syntactically least restrictive domain, this SP prediction would critically contrast with a UG theory of markedness, which would predict the same domain for anaphors to be also relevant for pronouns. As we have suggested above, it is thus only in the prediction for the pronoun domain that the SP proposal can be distinguished from a UG theory of markedness and where the empirical force of the SP proposal can best be evaluated. If markedness is defined in terms of extensional size of sets, not in terms of syntactic restrictiveness, GC(e) should be the child's unmarked hypothesis for pronouns.

Given these assumptions empirical evidence for the SP proposal could consist of several types:
1. Confirmation for the critical SP proposal that GC(e) is the unmarked hypothesis for pronouns would be found if behaviorally there were some stage in early acquisition, where the child prefers disjoint reference from all sentential antecedents for the pronoun. Disconfirmation for GC(e) as the unmarked domain on the other hand, would be

provided by data showing that at the earliest stages of acquisition, sentence-internal coreference to a c-commanding antecedent is generally allowed.

2. Confirmation of the SP proposal would exist if the ordering of children's hypotheses regarding GC for pronouns were as predicted by the SP, namely a domain associated to an extensionally smaller language before a domain associated to a larger language. Disconfirmation would be provided by the opposite ordering, e.g. GC(a) before GC(b—e).

3. In general, confirmation of the SP proposal for BT would exist if children's hypotheses for GC's for anaphors and pronouns were independent of one another.

6.1. *Empirical Studies*

In this section we consider results from a few selected studies which are representative of the general findings in the acquisition of lexical anaphors and pronouns. We will consider their bearing on the predictions assumed above. (We focus on published studies.) We begin with a foundational study on English-speaking children in this area.

6.1.1. *English*

Chien and Wexler (1987a) tested knowledge of locality conditions for anaphors and pronouns in 174 children ranging between the ages of 2;06 and 6;06. An act-out task (The 'Party Game') was used which required the children to follow instructions given by a puppet.[9] The instructions are exemplified in (23) and (24).

(23)   Reflexive:
    a. Kitty/Snoopy says that Sarah/Adam should give herself/himself a car.
    b. Kitty/Snoopy wants Sarah/Adam to give herself/himself a car.

(24)   Pronoun:
    a. Kitty/Snoopy says that Sarah/Adam should give her/him a car.
    b. Kitty/Snoopy wants Sarah/Adam to give her/him a car.

Figure 2 graphs correct responses (i.e., choice of correct antecedent, either long distance for pronoun or local for anaphor) by age group for reflexives and pronouns in the finite sentence examples (23a and 24a). Results for the infinitive complement examples (i.e., 23b and 24b) are similar. (Figure 2, like all following figures, is constructed by us to approximate the data reported in the published study, in this case, Chien and Wexler 1987a.)

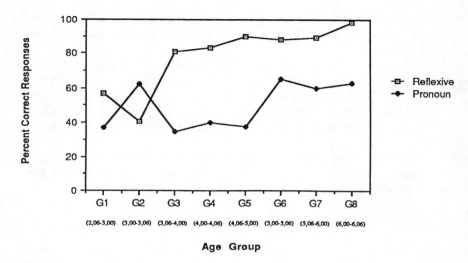

Fig. 2. English. Correct responses for reflexives and pronouns (Chien and Wexler 1987).

As Figure 2 shows, while the children's treatment of anaphors improves sharply at age 3;06 and then continues to improve steadily throughout the age groups, the development of correct treatment of pronouns is much slower and never reaches the success rate of the anaphors, even in the oldest age groups. The pronoun reaches its highest rate of accuracy in children between the ages of 5 and 6;06, where it is only about 65%.

Figure 3 graphs coreference between the pronoun and either a local or long-distance c-commanding antecedent. It thus displays the source of children's correct and incorrect responses on the pronoun.

As can be seen here, children allowed the pronoun to corefer locally throughout the different age groups. That is, they do not even impose disjoint reference in GC(a). In particular, the youngest children (G1—G3) freely allow this coreference, and while this error diminishes slightly for the older age groups (G7 & G8), it still has about a 40% occurrence rate.

These English data in general are not consistent with an hypothesis that children initially assume disjoint reference for pronouns everywhere. The initial hypothesis appears to be that pronouns are not obligatorily disjoint either locally or long distance. This hypothesis persists over development. In particular, there is no evidence that an extensionally smaller (subset) language for pronouns (e.g. GC(e)) is being chosen over a larger one (e.g. GC(a)). On the contrary, these data suggest that when children finally begin to assign disjoint reference to pronouns they do so in what corresponds to GC(a), i.e., the domain which is the extensionally largest language for pronouns, as defined by the SP. Thus, as the figure shows, errors in the local domain begin to wear off in the last 3 groups, impli-

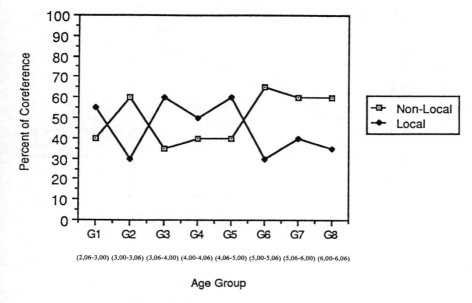

(2,06-3,00) (3,00-3,06) (3,06-4,00) (4,00-4,06) (4,06-5,00) (5,00-5,06) (5,06-6,00) (6,00-6,06)

Age Group

Fig. 3. English. Pronoun coreference (Chien and Wexler 1987).

cating GC(a). Since children also appear to evidence GC(a) for anaphors by group 3, this suggests that they hypothesize the same domain to be relevant for anaphors and pronouns.

6.1.2. *Dutch*

Results from studies on Dutch-speaking children corroborate these findings for English-speaking children (cf. Koster (1988) for an overview). Deutsch, Koster and Koster (1986), for example, tested 44 children between the ages of 4 and 10, using a picture-sentence matching task. Examples of the sentences are given in (25).

(25) a. De broer van Piet wast zich
         The brother of Piet washes himself

   b. De broer van Piet wast hem
       The brother of Piet washes him.

For each sentence the child was shown 4 pictures depicting two boys carrying out all possible versions of an activity. (Chance level was therefore at 25%.) Figure 4 gives the percentages of correct picture choice by age group for sentences with anaphors and pronouns.

Although these results cannot be compared directly to those in Chien

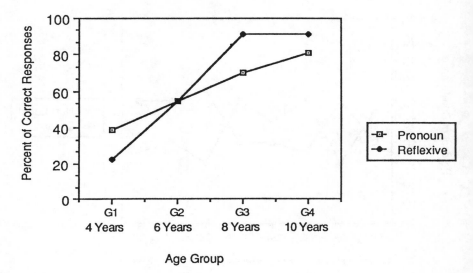

Fig. 4. *Dutch.* Correct responses for anaphors and pronouns (Deutsch, Koster and Koster 1986).

and Wexler's study because of the different experimental methods, they are similar in that they show a distinct lag in the development of the pronoun as opposed to that of the anaphor. Improvement for the pronoun is again less pronounced and success rate attained by the oldest group is significantly lower compared to that of the anaphor. More important to the evaluation of the SP proposal, however, is a fact pointed out by Koster (1988), namely that the most common error made by children in the DKK study was to assign to a pronoun a c-commanding local antecedent.

Thus, initial cross-linguistic evidence is consistent with the pattern observed in English acquisition, in that there is no evidence for wide-spread disjoint reference for the pronoun, but rather perseverance of an error of local coreference. By the same arguments above, GC(a) is the only domain evidenced for either pronouns or anaphors in Dutch as in English.

6.1.3. *Spanish*

Padilla-Rivera (1985, 1990) tested Spanish-speaking children on the knowledge of different locality conditions for anaphors and pronouns. In addition to sentences with anaphor and pronoun in the paradigm local position (i.e., similar to the earlier experiments) he used sentences where the anaphor and pronoun appear in the PP adjunct domain. Examples are given below.

(26)   Object position
   a. La hormiguita se cepilla detras de la zanahoria
       The ant-(dim) self brushes behind the carrot

   b. La rana la cepilla delante del helicoptero
       The frog 3 sg. combs in front of the helicopter

(27)   PP Adjunct
   a. La tortuga empuja la pelota al lado de si misma
       The turtle pushes the ball on side of herself

   b. La pantera patea el avioncito detras de ella
       The panther kicks the plane-(dim) behind her.

The PP adjunct domain is one in which anaphor and pronoun overlap in distribution, i.e., where they are not complementary. Since linguistically the PP domain (in Spanish as in English) allows either a free pronoun or anaphor to appear with coreference to the subject, it is of the type that has motivated linguists to propose distinct binding domains for anaphors and pronouns.[10] By studying children's acquisition of sentences like both (26) and (27) above, Padilla-Rivera was thus able to test whether children hypothesize identical binding domains for anaphors and pronouns, or whether they are aware of proposed different locality conditions which hold for the anaphor and the pronoun. In particular, if children do distinguish domains for anaphors (as involving main S) and for pronouns (as involving PP), then we would expect their behaviors to distinguish the anaphors and pronouns in the PP domain (27a and b) as in the S domain (26a and b). On the other hand, if GC(a) is the minimal syntactic domain for both anaphors and pronouns, then we may predict children to distinguish anaphors and pronouns first in the local GC(a) domain (e.g. 26a and b). If GC(e) is the first domain hypothesized for pronouns, in contrast to anaphors, then children should show disjoint reference on pronouns in both main clause and PP initially.

The experiment consisted of an act-out task in which 80 children, ranging between the ages of 3 and 9;11 used 4 puppets and several props to perform the sentences which were read out by the experimenter. Figures 5 and 6 give amounts of coreference and disjoint reference for pronouns and anaphors in both sentence types.

In general the results on the sentence types 26a, b are compatible with those of the previous studies in English and Dutch, in that they show 1) a slower development of correct treatment of the pronoun relative to the reflexive, and 2) perseverance of errors of local coreference for the pronoun. In addition, the data from the PP domain showed another domain where children do not initially assume disjoint reference everywhere for the pronoun, as would be predicted if they held an early GC(e) hypothesis for pronouns.

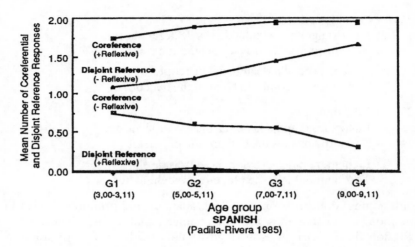

Fig. 5. Reflexives and pronouns in *sentence* object position.

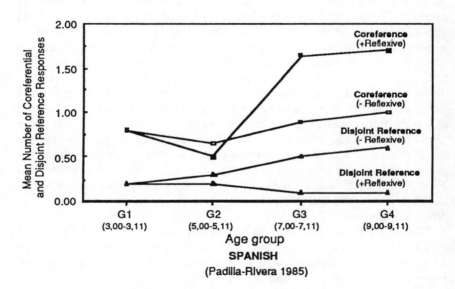

Fig. 6. Reflexives and pronouns in *PP* object position.

Notice that in Figure 5 (sentences 26a and b, S domain), the amount disjoint reference for the pronoun exceeds the amount coreference across all age groups. That is, when the pronoun appeared in the object position of a matrix sentence, children in Spanish tended to (correctly) prefer a sentence-external referent. Although there were errors offending local disjoint reference for the pronoun in Spanish, these appear to be eliminated more quickly in Spanish than in either Dutch or English. (Cf. Grimshaw and Rosen (1990) for a possible explanation of this fact in terms of Spanish clitics.) This could at first sight be taken as evidence for a preference for GC(e). Because the only domain tested in sentence types (26a) and (26b) was the S domain, disjoint reference here would be consistent with either GC(a) or GC(e) as the initial hypothesis for pronouns. Data on PP sentences (27a) and (27b), shown in Figure 6, however, show that GC(e) was not in fact the children's initial hypothesis. When given a choice between a correct sentence-external referent for disjoint reference (as GC(e) would require) and a correct sentence-internal referent, on the PP sentences, children preferred the latter. Thus, these data suggest again that children do not consider GC(e) to be the unmarked domain for pronouns. Again, the only evidence for pronoun is consistent with GC(a) as reflected in Figure 5. In addition, as the contrast between Figures 5 and 6 shows, the Spanish data suggest that children do not initially distinguish domains between anaphors and pronouns. That is, the behavior in S (Figure 5) does not simply replicate in the PP (Figure 6). Thus there is evidence that children do not assume independent domains for anaphors and pronouns as the SP proposal would appear to predict.

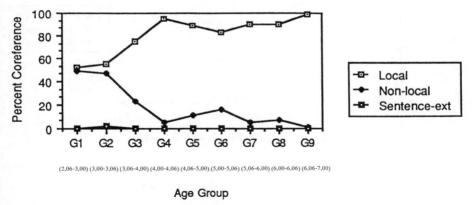

(2,06-3,00) (3,00-3,06) (3,06-4,00) (4,00-4,06) (4,06-5,00) (5,00-5,06) (5,06-6,00) (6,00-6,06) (6,06-7,00)

Age Group

Fig. 7. Chinese. Anaphor coreference (Chien and Wexler 1987).

6.1.4. *Chinese*

Acquisition data from Chinese are critical in the evaluation of the SP proposal, since this language has a reflexive (*ziji*) which can be long-distance bound in GC(e). As discussed in section 4.1.2., Chinese is a language where the choice of GC(e) for pronouns would not violate the Spanning Hypothesis. GC(e) could thus be reasonably expected to be the child's initial hypothesis, more so than for any other language we have reviewed here.

Significantly, even here the data suggest that children's initial hypothesis for pronouns is GC(a). Chien and Wexler (1987b) investigated the acquisition of locality conditions for the pronominals *ta* (− reflexive) and *ziji* (+ reflexive). They tested 150 children between the ages of 2;6 and 7;0 using the same experimental procedure as in their study of English-speaking children, i.e., the 'Party-Game', where the child was required to act out a sentence read out by a puppet. Examples are given below:

(28) a.   xiao-shizi yao xiaohua gei ziji yi-gen xiangjiao
            little lion want xiaohua give self one-CL banana

      b.   xiao-shizi yao   xiaohua gei   ta   yi-gen   xiangjiao
            little lion want xiaohua give her/him one-CL banana

Figures 7 and 8 plot percentage of amount coreference and disjoint reference (local, long distance and sentence-external) for the reflexive and the pronoun, respectively.

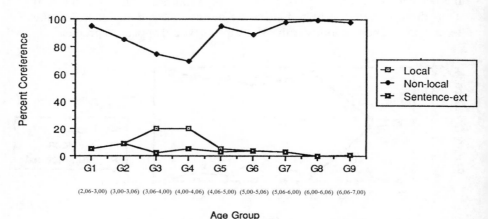

Fig. 8. *Chinese.* Pronoun coreference (Chien and Wexler 1987).

In general, we note again the slower development of the pronoun in terms of a lingering rate of error whereby local coreference was given for the pronoun. Note further that the response choosing a sentence-external referent for the pronoun, though correct, occurred very infrequently throughout the different age-groups, a result which again counters the SP prediction for disjoint reference everywhere as children's first, unmarked hypothesis for pronouns.[11] These results are critical, since, as we argued above, Chinese would clearly have allowed an extensionally unmarked hypothesis for pronouns, even given the 'Spanning Hypothesis'.

In conclusion, just as cross-linguistic variation in the domain for the pronoun does not cohere with the markedness prediction of the SP when anaphors appear to be set at GC(e), so these acquisition data show that children's initial assumptions about pronoun domain even in such languages does not expand to GC(e). In addition, the Chinese data on acquisition of anaphors suggest that although GC(e) would have been consistent with the SP for anaphors in this language, children do not evidence this grammatically non-minimal option.[12]

6.2. *Summary and Interpretation of Acquisition Studies*

Consider then how these results cohere with the Subset Principle as a principle of acquisition. Recall that what crucially distinguishes the SP proposal from a UG proposal is the prediction for the pronoun domain, since only here do intensional and extensional definitions of markedness differ. The results we have reviewed do not appear compatible with the SP predictions that the child's initial hypothesis for pronouns should be disjoint reference everywhere from c-commanding antecedents and that extensionally smaller language size determines the ordering of hypotheses. Acquisition results thus fail to confirm the essential empirical prediction of the SP, just as linguistic results do. Several results point to this conclusion.

First, cross-linguistic evidence shows that the initial stage of acquisition is marked by a common error for the pronoun where it is co-indexed to a local antecedent. The occurrence of this error in young children might be explained if we assume that the lexical distinction between anaphors and pronouns has not yet taken place (cf. Lust, Martohardjono, Mazuka and Yoon 1989, in preparation, for such a proposal.) However, we would at least expect that once this distinction has been made, the extensionally unmarked domains would be chosen for both anaphors and pronouns if the SP affects acquisition. The cross-linguistic acquisition data we have reviewed show a high success rate for anaphors at around age 4. Recall however, that this prediction is made by both the extensional SP proposal (albeit incidentally) and an intensional UG theory of markedness. Thus, this type of evidence cannot decide between the two theories.

Critically, the SP proposal would predict that the decrease of disjoint reference errors and high rate of local coindexing of GC(a) for the anaphor should be parallelled by a decrease in the local coreference error, and a preference for a sentence-external referent on the pronoun once pronouns and anaphors are distinguished in the lexicon. This, however, does not occur for the pronoun. Rather, in all the studies, children persisted in linking the pronoun locally to a c-commanding antecedent. Furthermore, when children were given the choice of a sentence-external referent, they tended to reject it, both when the pronoun occurred in sentence-object position (Chinese) and when it occurred in the PP adjunct position (Spanish). Thus, there is no evidence that at any stage in acquisition children are guided by an hypothesis that the domain for the pronoun is GC(e).

Second, when children finally reached a stage where they seemed to hold a hypothesis regarding the pronoun domain, this tended to be the syntactically smallest, but extensionally largest, domain available in the sentences. Thus, there is no evidence that the ordering of children's hypotheses for pronouns is guided by considerations of smallest language size.

Finally, children do not seem to freely assume independent and opposite domains for anaphor and pronoun, as has been hypothesized. Rather, initial evidence suggests that the child's early hypothesis attempts to link the definition of these domains (cf. Padilla-Rivera 1985, 1990).

7. CONCLUSIONS

The informal SP represents an important attempt to integrate issues of language 'learnability' and linguistic theory in a comprehensive learning model. It reflects an important attempt to formalize the notion 'most restrictive' hypothesis and thus to provide a mechanical definition of 'markedness'. The informal SP predicts that the markedness for pronoun domains would form a hierarchy opposite to that for anaphor domains. This follows because the markedness is determined extensionally, in terms of the size of the set of strings resulting under a particular choice of values. Thus GC(e) turns out to be the least marked domain value for pronouns. As we have seen, the critical empirical prediction which distinguished the SP proposal from a UG proposal regarding the initial unmarked domain for pronouns and anaphors (i.e., for the BT), was that these two should take opposite values made available by the proposed GC parameter, i.e., GC(a) and GC(e).

Results above have shown that this critical empirical prediction is not borne out by either cross-linguistic or acquisition data. These results have shown that attempts to account for this failure within the SP proposal by postulation of a Spanning Hypothesis are unsuccessful. One critical argument was based on the absence of any evidence (in either cross-linguistic variation or acquisition data) that GC(e) is the unmarked domain for pronouns even when the Spanning Hypothesis allowed it. Therefore both cross-linguistic data and acquisition data point to UG as the more plausible hypothesis.

We conclude that the Spanning Hypothesis is insufficient to account for the apparent synchronic and diachronic tendency for pronouns to be associated with what is, in Manzini's and Wexler's view, the most marked value of the GC parameter — GC(a), as well as for the evidence for an initial preference for GC(a) in language acquisition. Thus, a critical empirical prediction of this approach is disconfirmed. We believe that the failure of the predictions of markedness in the case of pronoun domains to conform to the observed cross-linguistic facts is a failure in principle, resulting from the position that markedness is determined extensionally, in terms of the size of the set of strings resulting under a particular choice of values.

These results lead us to conclude that an alternative research paradigm is motivated. This alternative pursues an intensional characterization of relative markedness of values for GC, under which a syntactically minimal domain, e.g. corresponding to GC(a), emerges as the unmarked value for both pronouns and anaphors — a claim which is more in accord with the cross-linguistic facts and the facts of acquisition. We suggest the possibility that at least in the particular case of domains for binding, and perhaps generally, the subset problem which M & W seek to address in their work may simply not arise in natural languages.

These results suggest that for locality conditions in general, including Principles A and B, the determination of the least marked domain has nothing to do with language size, but with the syntactic domain between the antecedent and the anaphor or pronoun. The least marked domain is the syntactically 'minimal' (cf. Chomsky 1985, 1986, and Rizzi 1990). (See Kapur et al. (1990) for more detailed discussion.)

This would appear to leave a learnability problem (a 'Subset Problem') unsolved in the case where GC's appear to vary. For example, consider the case where a learner is attempting to learn a domain for pronouns, and guesses A in (29), e.g. English, the syntactically most restricted, and therefore least marked domain, where the correct value is C, as in Icelandic.

(29) $[_C \ldots a2 \ldots [_A \ldots a1 \ldots b \ldots]]$

Changing the hypothesis from A to C in the case of a disjoint reference principle such as necessary for pronouns would require some form of evidence that pronoun b cannot be coindexed with antecedent a2. How could the correct grammar be achieved without recourse to negative evidence? One possibility is that, contrary to M & W's Lexical Parameterization Hypothesis, but consistent with the linguistic and acquisition evidence above, parameters are initially fixed for grammars, not lexical items. Specifically, if there is parametric variation in GC's, this could take place at the level of the definition of GC, and not in the definition of binding domains for pronouns and anaphors individually. The learner has a choice between, say, values, GC(a) and GC(c). Since both Principle A and Principle B make reference to GC, whichever value is chosen will be incorporated into both of these principles, as in a strong classic theory of

UG (e.g. Chomsky 1986). The expectation, then, consistent with the results we reviewed above, is that in a typical case the domains for pronouns and anaphors will converge. Then, positive evidence for determination of the value for GC can take the form of either anaphor interpretation or pronoun interpretation. Evidence for the pronominal domain then would have consequences for the anaphoric domain, and the converse. Note that such linking of values is not wholly avoided in M & W's hypothesis, since this is the effect of the Spanning Hypothesis (cf. also Vikner (1985, p. 47) for a suggestion along similar lines). The Icelandic learner, for example, could initially assume GC(a) (for both pronouns and anaphors) but be led to GC(c) by instances of anaphors bound in GC(c), i.e., positive evidence.

We do not suggest this scenario as a veridical proposal (see Lust, Martohardjono, Mazuka and Yoon 1989), but only to show that even in cases where the subset problem might seem most likely to arise, it need not, depending on one's analysis. As M & W note, the problem does not arise with respect to most of the other parameters suggested in the linguistic literature, whose values do not generate languages which are in a subset relation.

These results cohere with Chomsky's (1986) E-language/I-language distinction. The failure of the proposed learning module, and its associated SP, to sufficiently or accurately predict either cross-linguistic or first language acquisition restrictions on anaphora is in fact a failure of an E-language principle (which computes size of sets of sentences). The fact that the most restrictive local binding domain, defined grammatically, appears to significantly constrain both the possibilities for cross-linguistic variation in pronoun and anaphor locality and the cross-linguistic first language acquisition data regarding pronouns and anaphors, regardless of the extensional size of the corresponding language, is in accord with an intensional representation of this area of language knowledge. These results are important because they suggest that an explanation for cross-linguistic variation (in the case of BT) results from a strong theory of UG. They are also important because they suggest that the nature of the data which constitute evidence in natural language acquisition must be at least in part deductively derived.

APPENDIX
THE DEVELOPMENT OF DISJOINT REFERENCE IN GERMANIC
(HARBERT 1990)

Our claim is that in Gothic (and Germanic), pronouns had to be free within the domain of the minimal finite clause containing them. It was not sufficient that they be free just in the domain of a subject. That is, in the terms of M & W's GC hierarchy, pronouns observed GC(c), not GC(a). This is reflected, e.g. in the asterisk in (1a).

(1) a.   jah bedun ina*j* *allai gaujans* . . . [*s* PRO*j* galeiþan fairra *sis/(*ina)*]
          and asked him all of the inhabitants*i* to go(Inf) from selves*i*/*them*i*

          "And all of the inhabitants asked him to go from them" (Luk 8:37) (Greek
          model has a pronoun)

   b.   (*eis*) bedun ina ei   [*s* (is) uslaubidedi *im/(*sis*) . . . galeiþan]
          They asked him that   he allow(Fin)   *them/*selves* to go

          "They asked that he allow them to go." (Luk 8:32)

Of course, since Gothic is an extinct language, the claims represented in (1) do not reflect
direct judgments of native speakers. We believe that the distribution of asterisks presented
here may nonetheless be reconstructed with considerable certainty; the Gothic corpus
consists in the main of a translation of the Greek New Testament which is, in accordance
with the conventions for translation of the time, exceedingly faithful and literal, giving no
evidence of a tendency to deviate gratuitously from the model. Yet in the numerous
constructions of the type in (1a) where the Greek text uses a pronoun with intended
coreference with the matrix subject, across the boundary of a nonfinite clause (infinitival or
participial), the Gothic text **systematically** translates that pronoun not with a Gothic
pronoun, but with a reflexive. (See Harbert (1978) and later work for extensive documentation of this.) (The Greek model has pronouns in such cases because the reflexive in
New Testament Greek, as in English, observed GC(a).) The systematic translation of
Greek pronouns as reflexives here demonstrates not only that long binding of the reflexive
to the matrix subject was possible in such constructions in Gothic, unlike Greek, but also
that pronouns in such positions were interpreted as necessarily disjoint in reference with
the matrix subject. In the absence of the latter assumption, it is not clear how the observed
systematic deviation from the Greek model in such cases could be accounted for. In other
contexts, including cases of finite complements such as (1b), Greek pronouns are translated as pronouns in Gothic — never as reflexives. As was pointed out in Harbert (1986a),
the claim that in Gothic pronouns must be free in the domain of the minimal finite clause
(GC(c)) also accounts for a second construction with respect to which the Gothic text
diverges systematically from the Greek model for translation by using a reflexive instead of
a pronoun. This is represented by (2).

(2)   jah (*is*) qaþ   im   in [laiseinai *seinai/(*is)*]
        and he*i* spoke to-them in teaching self's*i*/*his*i*

        "And he spoke to them in his doctrine." (Mk. 4:2)
        (Greek text has pronoun)

     In Greek, as in English, pronominal possessives (his/autou) could be used as subjects of
NP's with intended coreference with the subject of the containing clause (and usually were,
the possessive reflexive being reserved for emphatic uses). This can be accounted for again
by assuming that these languages observe GC(a). In Gothic, however, these pronouns too
are systematically translated as reflexives — as we might expect if the presence of a subject
within NP was not sufficient in that language to define NP as a GC for pronouns.
     That the distribution of forms in Gothic is to be accounted for in this way is strongly
suggested by the fact that it is quite similar to that found in Modern Icelandic, where
subject nominals also fail to create opacity either for the reflexive *sig/sin* or for pronouns
in constructions like (1a) or (2). Thus, in Icelandic, a pronoun in the complement in a
sentence like (1a) must be, as in Gothic, disjoint in reference with the main clause subject.
The behavior of pronouns in Icelandic indicates at least that the restrictions on pronoun
interpretation claimed for Gothic are possible ones.
     In other Germanic languages, including English and German of the West Germanic

subgroup and some dialects of Danish, pronouns in constructions like (1a) **may** corefer with the subject of the main clause (while reflexives may not). Thus, both patterns are found together in Germanic. That the Gothic/Icelandic pattern is the historically prior one, and the one to be reconstructed for the parent language, is indicated by comparative considerations, as well as by evidence of historical developments within languages of the other type. Thus, in Old and Middle High German, reflexives were used (exclusively, so far as we can determine) in cases like (3) to refer to the matrix subject, while in Modern German the reflexive cannot be so used, whereas the pronoun can. This indicates a change away from an original Gothic-type pattern in the history of German.

(3) a.   Uote$_i$ bat dô drâte die boten$_j$ [PRO$_j$ für sich$_i$ gên] (MHG — Nib 772:1)
           Ute$_i$ asked then quickly the messengers before self$_i$ to-go

    b.   Ute$_i$ bat die Boten$_j$ [PRO$_j$ vor ihr$_i$/*sich$_i$ her zu gehen] (Mod Ger)
           Ute asked the messengers before her/*self to go

Similarly, reflexives are almost always found in constructions like (1a) in Old Norse (though Falk and Torp (1900) and Nygaard (1905) note the existence of sporadic exceptions). Thus, for example, Falk and Torp state (p. 133) "with the infinitive *sinn* is ordinarily used [in ON] referring to the sentential subject," as in:

(4) a.   þórvaldr bað byskup fara til Islands með sér ok skíra foður sinn
           þorvaldr asked bishop to go to Iceland with self and baptize father self's

That is, it appears that in Old Norse the reflexive was required here, as in Modern Icelandic. Falk and Torp note, however, that in Danish and in Dano-Norwegian, pronouns are now generally used in such constructions,

(4) b.   hun$_i$ bad ham give hende$_i$ tid
           she asked him to give her time

though many dialects (especially in Dano-Norwegian) still use here what they refer to as the "old" *sig*. (In fact, as reported in Harbert (1983) some speakers of Danish still appear not to allow pronouns in this context.) Similarly, Iversen (1918), in his syntax of the Tromsø dialect of Bymaal (Dano-Norwegian), notes the frequent use there as well of the pronoun, rather than the reflexive in such constructions (though the reflexive is possible there as well, with the distribution determined largely by pragmatic factors), but characterizes the reflexive as the normal form in such constructions at earlier stages.

Thus, the Gothic pattern — reflexives but not pronouns with reference to the main clause subject — is directly evidenced in Gothic, Icelandic, and some dialects of Danish, and supported by indirect evidence in Middle High German and Old Norse, where proximate pronouns never or at most very rarely are found to occur in such constructions. Moreover, in the case of many of the languages exhibiting the English-type pattern (pronouns allowed with reference to the matrix subject), e.g. in German, Danish and Dano-Norwegian, there is evidence that this pattern is a result of a historical development.

NOTES

[1] This paper is prepared with the partial support of NSF under grant #MIP-8602256 to Gianfranco Bilardi and grant #8318983 to Barbara Lust, and SSHRC of Canada under grant #452896086 to Gita Martohardjono. We thank Jim Gair, Jim Huang, Alice Davison, M. Everaert, G. Longobardi, Rita Manzini, Ken Wexler, Gianfranco Bilardi, Virginia Valian, Fred Landman and Kashi Wali among others, for discussion of the issues raised in this paper.

² It is not clear to us whether M & W intend that the particular values on the hierarchy are provided independently by the theory of grammar or whether they arise solely from learnability theory — i.e., the subset calculation. See Kapur et al. (1990) for discussion of the implications of both of these interpretations.

³ It is not clear to us how the second half of this principle, as formulated, is satisfied in a language like Icelandic, for example, where pronouns appear to be associated with value (c) of the governing category parameter and value (a) of the proper antecedent parameter, in spite of the discussion in Manzini and Wexler (1988, p. 438). Consider the case of the pronoun in the following Icelandic sentence:

(i)   Ég tel   Jón   hafa   lofað   Bill$_i$ [að PRO$_j$ heimsaikja hann$_i$]
       I   believe John to have promised Bill$_i$ to   visit   him$_i$

Here, the pronoun is neither free in its unmarked GC, the root clause, (since it is (object) bound in the intermediate clause) nor free from its unmarked proper antecedents (since it is bound to an object). We suspect therefore that the formulation in (5) is insufficiently precise to convey what Manzini and Wexler actually intend.

⁴ The stipulation in (5) appears, in effect, to 'override' independent parameter settings in particular instances.

⁵ In particular, in Yoruba, the same 'pronoun' appears obligatorily in cases of subject WH-movement, showing that at least sometimes it is associated with an operator-bound variable. Moreover, it appears that this clitic can indeed corefer with the higher subject in a configuration like (i), just in case B is a phrase which is also an island for WH-movement — e.g., a manner clause.

(i)   NP$_i$ ... [$_B$ ... clitic$_i$ ...]

As Pulleyblank points out, this parallel between the domain of WH-movement and the domain of disjoint reference assignment to the clitic is captured under the assumption that the clitic is indeed associated with a variable, which must, by Principle C, be free in the domain of the operator that binds it. See Pulleyblank (1986) for further arguments.

⁶ Note, for instance, that they have inherent semantic content beyond that of the antecedent. Other possible instances of long distance disjoint pronouns in the literature which we have noticed include the Tamil form cited by Yadurajan (1989, p. 195). (We thank Kashi Wali for bringing this to our attention.) The examples he gives, however, seem again to be subject to an alternative interpretation; we may be dealing here not with binding-theoretic effects, but with the pragmatic effects that typically determine whether overt or null pronouns are used in pro-drop languages, and related obviation effects (cf. Montalbetti 1984, p. 85).

⁷ We claim in the appendix to this paper that Gothic is another such instance of the putative GC(c). G. Longobardi has suggested to us (personal communication) that in view of the near universal preference for GC(a) as the value for pronouns, perhaps there is no parameterization for pronominal governing categories at all. If that were the case, of course, then there would be no question of extensional determination of domain size for pronouns, and the SP would be completely otiose for pronouns. It is not clear how the facts of Gothic and Icelandic could be accounted for under such a view, however. Finally, we thank M. Everaert (personal communication) for bringing to our attention the facts in (i), which suggest that GC(c) is in fact not an entirely accurate characterization of the pronominal domain in Icelandic. For further discussion of these facts, see Kapur et al. 1990.

(i)   Jón$_i$ heyrði lysingar   Maríu á honum$_i$
       John heard descriptions Mary's of him

⁸ As we discuss further in Lust, Martohardjono, Mazuka and Yoon (1989) in order for

the SP hierarchies to act as an efficient guide in constraining the child's hypotheses regarding locality, the child must have previously distinguished lexical pronominals in his/her lexicon as either anaphors or pronouns. Without this initial step the child could not know which hierarchy is relevant to a lexical item. Alternatively, it could choose the wrong hierarchy for a given item. For example, if the child hypothesizes a lexical pronominal to be an anaphor, when in fact it is a pronoun, this would lead her to choose GC(a) as the relevant domain when in fact the SP determines GC(e) as the unmarked domain for this situation. Given the SP as the sole determiner of learning, it would then be unclear whether and how the child could retreat from this error. The problem of differentiating pronouns and anaphors prior to applying the locality conditions is not unique to the SP, since the child must determine whether a pronominal is governed by Principle A or B, i.e., whether it is to be bound or free in a domain. However, the SP compounds the problem by adding the requirement of selecting different unmarked domains for anaphors and pronouns.

[9] Chien and Wexler (1987) actually reports several studies on anaphora. Here, we only review the one which is most relevant to our discussion.

[10] Huang (1983) formulates this distinction in terms of the additional notion of "accessible subject" for anaphors but not pronouns. On this view, the PP would be a Binding Domain for the pronoun, resulting in the fact that it is "free", but would not be a Binding Domain for the anaphor, reflecting the fact that it is still "bound" to the subject of the matrix clause.

[11] It should be noted that the children in the Chinese study appeared to make fewer errors on the pronoun compared to the children in the other studies of English and Dutch. They assigned more disjoint reference in the local domain than children in either of those languages.

[12] Interestingly, the adults used in this study as controls are also reported to have preferred GC(a) to GC(e) for the anaphor. This result too is in accord with an intensional UG determination of markedness.

REFERENCES

Angluin, Dana: 1980, 'Inductive Inference of Formal Languages from Positive Data', *Information and Control* **48**, 117—135.

Berwick, Robert: 1982, *Locality Principles and the Acquisition of Syntactic Knowledge*, unpublished doctoral dissertation, Department of Electrical Engineering and Computer Science, MIT.

Berwick, Robert: 1985, *The Acquisition of Syntactic Knowledge*, MIT, Cambridge.

Bresnan, Joan, Per-Kristian Halvorssen and Joan Maling: 1983, 'Invariants of Anaphoric Binding Systems', paper presented at Cornell University, Ithaca, New York.

Chien, Yu-Chin and Ken Wexler: 1987a, 'Children's Acquisition of the Locality Condition for Reflexives and Pronouns', *PRCLD, 26*, Stanford.

Chien, Yu-Chin and Ken Wexler: 1987b, 'Comparison Between Chinese-speaking and English-speaking Children's Acquisition of Reflexives and Pronouns', paper presented at the 12th Boston University Conference on Language Development, October, 1987.

Chomsky, Noam: 1981a, *Lectures on Government and Binding*, Foris, Dordrecht.

Chomsky, Noam: 1981b, 'Markedness and Core Grammar', in Adriana Belletti, Luciana Brandi and Luigi Rizzi (eds.), *Theory of Markedness in Generative Grammar*, Scuola Normale Superiore di Pisa, Pisa, 123—146.

Chomsky, Noam: 1986, *Knowledge of Language. Its Nature, Origin, and Use*, Praeger, New York.

Deutsch, Werner, Charlotte Koster and Jan Koster: 1986, 'What Can We Learn from Children's Errors in Understanding Anaphora?', *Linguistics* **24**, 203—225.

Echavarri-Dailey, Anna: 1986, 'Basque and Binding Theory', unpublished manuscript, Cornell University.

Falk, Hjalmar and Alf Torp: 1900, *Dansk-Norskens Syntax*, H. Aschehoug, Kristiania.

Faltz, Leonard: 1977, *Reflexivization: A Study in Universal Grammar*, Ph.D. dissertation, University of California.

Fisher, Karen: 1984, 'The Syntax and Semantics of Anaphora in Khmer', Master's thesis, Cornell University.

Gair, James: 1988, 'Kinds of Markedness', in Susanne Flynn and Wayne O'Neil (eds.), *Linguistic Theory in Second Language Acquisition*, pp. 225—250.

Grimshaw, Jane and Sarah Rosen: 1990, 'Knowledge and Obedience: The Developmental Status of the Binding Theory', *Linguistic Inquiry* **21**(2), 187—223.

Harbert, Wayne: 1978, *Gothic Syntax: A Relational Grammar*, Ph.D. dissertation, University of Illinois.

Harbert, Wayne: 1983, 'Germanic Reflexives and the Implementation of Binding Conditions', in Irmengaard Rauch and Gerald Carr (eds.), *Language Change*, Indiana University Press, Bloomington, 89—127.

Harbert, Wayne: 1986a, 'Markedness and the Bindability of Subject of NP', in Fred Eckman, Edith Moravcsik and Jessica Wirth (eds.), *Markedness*, Plenum Press, New York, 139—154.

Harbert, Wayne: 1986b, *Binding, SUBJECT and Accessibility*, in Robert Freidin (ed.), *Principles and Parameters in Comparative Grammar* 1990, MIT Press, 29—55.

Huang, C.-T. James: 1982, *Logical Relations in Chinese and the Theory of Grammar*, unpublished Doctoral dissertation, MIT.

Huang, C.-T. James: 1983, 'A Note on the Binding Theory', *Linguistic Inquiry* **14**, 554—561.

Iversen, Ragnvaid: 1918, *Syntaksen i Tromsø Bymaal*, Bymaals-Lagets Forlag, Kristiania.

Jakubowicz, Celia: 1984, *On Markedness and Binding Principles*, in Carl L. Baker and John McCarthy (eds.), Proceedings of the fourteenth North Eastern Linguistic Society, University of Massachusetts, Amherst, pp. 154—228.

Kapur, Shyam, Barbara Lust, Wayne Harbert and Gita Martohardjono: 1990, *On Relating Universal Grammar and Learnability Theory: Intensional and Extensional Principles in the Representation and Acquisition of Binding Domains*, in preparation.

Koster, Jan: 1981, 'Configurational Grammar', in Adriana Belletti, Luciana Brandi and Luigi Rizzi (eds.), *Theory of Markedness in Generative Grammar*, Scuola Normale Superiore di Pisa, Pisa.

Koster, Charlotte: 1988, 'An Across-Experiments Analysis of Children's Anaphor Errors', in Ger de Haan and Wim Zonneveld (eds.), *Formal Parameters of Generative Grammar, OTS Yearbook IV*, ICG, Dordrecht, pp. 53—74.

Lee, Chungmin: 1988, 'Issues in Korean Anaphora', in Eung-Jin Baek (ed.), papers from the Sixth International Conference on Korean Linguistics, University of Toronto, pp. 339—358.

Lust, Barbara, Reiko Mazuka, Gita Martohardjono and Jeong-Me Yoon: 1989, *On Parameter Setting in First Language Acquisiton: The Case of Binding Theory*, presented at GLOW, The Netherlands, April 1989, in preparation.

Manzini, Rita M. and Ken Wexler: 1987, 'Parameters, Binding Theory and Learnability', *Linguistic Inquiry* **18**(3), 413—444.

Mohanan, K. P.: 1982, 'Grammatical Relations and Anaphora in Malayalam', *MIT Working Papers in Linguistics, 4*, pp. 163—190.

Montalbetti, Mario: 1984, *After Binding: On the Interpretation of Pronouns*, unpublished Doctoral dissertation, MIT.

Nygaard, Marius: 1905, *Norrøn Syntax*, H. Aschehoug, Kristiania.

Oshima, Shin: 1979, 'Conditions on Rules: Anaphora in Japanese', in George Bedell, Eichi

Kobayashi and Masatake Muraki (eds.), *Explorations in Linguistics: Papers in Honor of Kazuko Inoue.*

Padilla-Rivera, Jose: 1985, *On the Definition of Binding in Spanish: The Roles of the Binding Theory Module and the Lexicon*, Doctoral thesis, Cornell University, Ithaca, New York.

Padilla-Rivera, Jose: 1990, *On the Definition of Binding Domains in Spanish: Evidence from Child Language*, Kluwer, Dordrecht.

Pulleybank, Douglas: 1985, 'Clitics in Yoruba', in H. Borer (ed.), *Syntax and Semantics, 19*, Academic Press, Harcourt Brace Jovanovich, Orlando, 43—64.

Rizzi, Luigi: 1990, *Relativized Minimality*, MIT, Cambridge.

Saxon, Leslie: 1983, 'Disjoint Reference Between Anaphor and Antecedent', paper presented at the LSA annual meeting.

Thráinsson, Häskuldur: 1976, 'Some Arguments Against the Interpretive Theory of Pronouns and Reflexives', *Harvard Studies in Syntax and Semantics* **2**, 573—624.

Vikner, Sten: 1985, *Parameters of Binder and Binding Category in Danish*, MA thesis, Geneva, Switzerland.

Yadurajan, K. S.: 1988, 'Binding Theory and Reflexives in Dravidian', in Veneeta Srivastav, James Gair and Kashi Wali (eds.), *Cornell University Working Papers in Linguistics, 8*, A Special Issue: Papers on South Asian Linguistics, 181—203.

Yokoyama, Olga: 1980, 'Studies in Russian Functional Syntax', in Susuma Kuno (ed.), *Harvard Studies in Syntax and Semantics III*, Department of Linguistics, Harvard University, 191—208.

KENNETH WEXLER

THE SUBSET PRINCIPLE IS AN
INTENSIONAL PRINCIPLE*

The fundamental problem of modern linguistic theory is the problem of explanatory adequacy, or learnability. Namely, how does the child construct her grammar from the input (the environment)? Extensive research has shown that the input greatly underdetermines the grammar. The general answer is that there are innate principles of grammar, and that a small amount of knowledge, particularly the lexicon and the values of grammatical parameters, is the result of experience.

The ordinary way in which linguistic experience has an effect is that learned knowledge must be consistent with the experience. Such an assumption is made, implicitly or explicitly, by just about every particular approach in the field.[1] The nature of the experience might be limited. Thus it is generally agreed that the child does not make use of negative evidence. UG is strong enough that positive evidence is all that is needed for what learning takes place.

The lack of negative evidence has led to a well-known question concerning how learning takes place. Suppose, for the sake of simplicity, that there are two possible values that a piece of learning may result in, values A and B. If it turns out that every structure that is grammatical under value A is also grammatical under option B (but not vice versa), then if the learner ever decides that option B is correct, under the no-negative evidence assumption, there is no way to ever retreat to option A, because all future evidence will be consistent with both option A and option B.[2]

The well-known solution to this problem is to assume the *Subset Principle*, which says that, if every structure that is grammatical under A is also grammatical under B, then choose option A if it is consistent with the input. The Subset Principle was formalized in Manzini and Wexler (1987) and Wexler and Manzini (1987); there is no need to repeat that formalization here, since the intuitive idea is obvious. We will provide alternative formalizations as we proceed. Manzini and Wexler and Wexler and Manzini argued that the Subset Principle could be taken to be an independent principle, in effect part of a learning module, sharpening the consistency assumption, which was already (in effect) part of a learning module. It would thus allow UG to not have to state a markedness hierarchy, a hierarchy which would be unmotivated in linguistic theory itself. In this way it cast a new light on the concept of markedness, providing the concept of markedness with a deductive structure, rather than making it simply a stipulative concept.[3]

217

Eric Reuland and Werner Abraham (eds), Knowledge and Language, Volume I, From Orwell's Problem to Plato's Problem: 217–239.
© 1993 *by Kluwer Academic Publishers. Printed in the Netherlands.*

   The Subset Principle has provided a basis for a number of researches in linguistic theory and in language acquisition. As with the introduction of many new concepts,[4] confusion sometimes arises. This confusion is especially possible if the foundations of generative grammar are not taken into account, but rather an older, descriptive framework is assumed. The Subset Principle, simple as it is, is inextricably linked to a view of linguistic theory[5] in which the central problem is taken to be learnability, the problem of the poverty of the stimulus.[6] The problem of learnability played no role in the older, descriptive frameworks. Thus, within that tradition the Subset Principle would not have a natural role to play. Rather, stipulative markedness hierarchies might be assumed, the conceptual status of which was often less than clear, as we will discuss. At any rate, markedness hierarchies of this kind have nothing to do with the fundamental problem of linguistic theory. The purpose of the present paper is to attempt to dispel the confusions that have arisen when the descriptive, non-explanatory-adequacy-oriented tradition attempts to confront the Subset Principle.

   Kapur, Lust, Harbert and Martonhardjono (1991), henceforth KLHM, have written most extensively in this vein about the Subset Principle. They have produced a number of arguments against the Subset Principle. I will show that these arguments are either incoherent or theoretically or empirically flawed. In addition I will attempt to show how the fundamental motivation for their arguments comes from a framework in which explanatory adequacy (the learnability problem) is not a primary concern. Basically the authors ignore what Chomsky (1986a) has called "Plato's Problem." Interestingly, KLHM's arguments also use language in a particular way which helps to create confusion about some very simple concepts. In a small way, this is a good illustration of the other major problem that Chomsky (1986a) discusses — "Orwell's Problem." (After we discuss the failures of KLHM's arguments it will become apparent how their particular use of language helps to throw light on Orwell's problem. In particular, the language that KLHM use is unjustified by the concepts under discussion.)

1. THE SUBSET PRINCIPLE AS AN INTENSIONAL PRINCIPLE

As Chomsky (1986a) points out, grammatical principles are *intensional*, not *extensional* principles. That is, they are part of the *I-language*, not the *E-language*. The I-language is a set of principles and representations in the mind, including Universal Grammar and some properties set by experience. Chomsky in fact argues that there may not even be a coherent scientific notion of *E-language*.

   The Subset Principle is clearly an I-language (intensional) principle. It relates the properties of 2 settings of parameters.[7] Perhaps the statement of the Subset Principle in terms of sets obscures this fact, but that is only a notational convenience.[8] Let us state the Subset Principle (SP) as in (1).

(1)   *Subset Principle*: Suppose parameter P has 2 values, i and j, such that for all derivations D, if D is grammatical under setting i then D is grammatical under setting j. Then value i is unmarked with respect to value j.

If one value of a parameter is unmarked with respect to another value, the learner will select the unmarked value, if the data do not contradict that choice. This is stated in (2).

(2)   *Markedness*: If (i) value i of parameter P is unmarked with respect to value j, and (ii) the primary data (input) is consistent with value i, then the learner selects value i of parameter P.

Derivations are a property of the I-language, not the E-language (if there even is such a thing). Clearly (1) is an intensional principle. And clearly (1) is equivalent to the Subset Principle as stated, e.g. in Manzini and Wexler (1987).

The set of derivations is a property intrinsic to linguistic theory, and part of the I-language. For example, we say that a sentence S is *ungram-matical* if and only if all grammatical derivations do not generate S.

Chomsky (class lectures, Fall 1990) has argued that there is no scientific notion of *set of grammatical sentences*. He has argued that strings (derivations) are what they are, that they have all sorts of markings. It has become customary to say that a string (derivation) is grammatical if and only if it hasn't violated any property of grammar. However, Chomsky argues that there is no reason to distinguish the class of grammatical strings (derivations) from its complement, which contains derivations with at least one violating property. Derivations with violations are still part of the I-language. Let us call this assumption the property of "no distinguished grammatical class" (*ndgc*).

Suppose that *ndgc* is true. It is still possible (and quite natural) to define the subset Principle, in a precise way, as an intensional principle. Basically the idea is to relativize the relationship of the derivations to violations of a particular grammatical principle. For a parameter P, for any derivation D, let D(i) be the derivation carried out under value i of the parameter.

(3)   *Subset Principle (Relativized Statement)*: Suppose that principle X allows for 2 values, i and j, of a parameter P. Suppose furthermore that for all derivations D, if D(j) violates X, then D(i) violates X. Then value i of parameter P is unmarked with respect to value j.

To take an example, suppose, following Manzini and Wexler, that the governing category of an anaphor is parameterized. For simplicity, consider just two 2 values of the parameter, value (a), the domain of the closest subject,[9] a small domain, and value (d), the domain of the closest finite tense. Principle A says that the anaphor must be bound in its governing category. Suppose that derivation D(d) carried out under value d violates Principle A. This means that an anaphor is not bound within the domain of finite tense. This implies that the anaphor is not bound within the domain of the closest subject. Thus the derivation D(a) carried out under value (a) of the parameter will violate Principle A. This result is true for all derivations D. Thus, by (3), value (a) is unmarked with respect to value (d).

It is clear that, at least for the cases considered to date, the Relativized Statement of the Subset Principle (3) is equivalent to the Subset Principle stated in other forms, e.g. (1), or the statement in terms of sets in Manzini and Wexler.[10] What it depends on is the clear relation between the setting of a parameter and the effect on the derivations carried out under a particular principle. The Independence Principle of Manzini and Wexler (1987) and Wexler and Manzini (1987) is formulated to make these relations clear.

Note that what is particularly important in the ability to state a principle like (3) is a strong theory of UG. If the grammatical setting of the Subset Principle is not highly constrained, then the Subset Principle will not apply. In this regard note that, to my knowledge, all the cases in which the Subset Principle has been applied to date are such that the parameter in question is only relevant to one principle. Although all the cases haven't been worked out, it seems to me that even in cases where a parameter is relevant to more than one principle the Subset Principle is defined appropriately in (3). What might happen in such a case is that it will be more difficult for the Subset Principle to apply. If it turns out that there are important cases in which a parameter is relevant to more than one principle, then further analysis of such cases might be warranted.

The important point for present concerns is that the Subset Principle in (3) is an intensional principle, and furthermore, an intensional principle under a very strong conception of I-language, in which there is no distinguished set of grammatical sentences. Furthermore, (3) applies correctly to all Subset Principle cases considered to date, and is likely to apply correctly to all potential Subset Principles cases. The formulations given here make even clearer what was already known previously, that the Subset Principle is an intensional principle. Thus there is no argument against the Subset Principle on conceptual grounds.

2. STIPULATIVE MARKEDNESS

KLHM proposes an alternative to the Manzini and Wexler Subset Principle analysis of governing categories. We will argue that the Subset

Principle analysis of binding theory variations is distinctly superior to the alternative analysis that KLHM propose.[11] We should note, however, that a different analysis of a particular domain, for example binding theory, does not show that the Subset Principle is incorrect. At most it can take away some of the motivation for the Principle.

Manzini and Wexler argue that the governing category parameter also is relevant for pronouns, and that the setting of this parameter for pronouns can be different from the setting for the parameter for anaphors. Thus the domain in which a pronoun must be free (Principle B) may be different in a particular language from the domain in which an anaphor must be bound (Principle A). This is allowed by the *Lexical Parameterization Principle*, which ties parameters to lexical items. Further, Manzini and Wexler show how the Subset Principle implies that the markedness hierarchy for pronouns is opposite to the markedness hierarchy for anaphors. This is because Principle A requires that an anaphor be bound, whereas Principle B requires that a pronoun be free. The Relativized Statement of the Subset Principle (3) leads to exactly the same conclusion.

KLHM accept Manzini and Wexler's claim that variation of governing category is possible for pronouns as well as anaphors and that there is a markedness hierarchy for the parameter values. However, they reject the Subset Principle. Rather, they claim that the "syntactically minimal domain"[12] is the unmarked value of the parameter for both anaphors and pronouns. They say that this is a "strong UG" view.

Soon we will turn to the evidence concerning which is the correct theory of markedness. First let us deal with the conceptual issues. I see no reason to call the KLHM theory a "strong UG" view of markedness. What this theory does is to stipulate markedness, so we might as well call it a theory of "stipulative markedness" (SM). Linguistic theory in its own terms has no reasons to provide a markedness theory for the values of parameters. The statement of a markedness hierarchy is simply a stipulation which reduces the conceptual simplicity and power of the theory.

Consider a well-known analogy. Stowell (1981) argued that phrase-structure rules don't exist. He showed how many cases which appeared at first sight to need phrase-structure rules could be explained by the interaction of general principles. This was considered a major advance in the development of syntactic theory because a module of complex rules was no longer needed. The theory of Stipulative Markedness, it seems to me, has the same status as the theory of phrase-structure rules. If general principles (the Subset Principle) can be shown to predict the correct markedness hierarchies, then we simply don't need the theory of Stipulative Markedness. In my view, this is exactly the situation, at least with regard to the cases under discussion and a number of other cases. The ideal theory would be one in which we didn't need stipulative markedness at all. Of course it remains to be seen whether stipulative markedness can be completely done away with, but I think that the case is clear that it's not necessary for binding theory. At any rate, conceptually, stipulative Markedness is a retrogression from a principled theory, a retrogression that empirical arguments might push us towards, but definitely not the desired state.

The Subset Principle is an attempt to extend the spirit of a principles and parameters system of UG to the question of markedness hierarchies. It attempts to provide a principled account of the hierarchies and thus not depend on stipulated hierarchies (analogous to complex stipulated rules of phrase-structure). Note that it locates the hierarchies not in one stipulative statement, but makes them result from the interaction of principles which are in separate modules (e.g. the Binding Principles, together with the Subset Principle). See Wexler and Manzini (1987) and Safir (1987).

KLHM also argue that Stipulative Markedness is an intensional concept whereas the Subset Principle is extensional. But we have already seen in section 1 that the Subset Principle is intensional. Thus there is no difference in this regard between the two conceptions of markedness.

3. CROSS-LINGUISTIC DATA ON PRONOUN GOVERNING CATEGORIES

KLHM's major empirical argument against the Subset Principle is the following. According to Manzini and Wexler (1987) and Wexler and Manzini (1987), the Subset Principle predicts that the unmarked governing category for the pronoun is the largest (root) governing category. In terms of (3), if a pronoun violates Principle B when the governing category is taken to be a small domain, say the domain of finite tense, then the pronoun will violate Principle B if the governing category is taken to be the root sentence. Thus, by (3), the domain of the root sentence is unmarked with respect to the domain of finite tense. Identical arguments show that for the pronoun (i.e., for Principle B), the root governing category is unmarked with respect to all other governing categories. (In general, the markedness hierarchy turns out to be exactly opposite to that for the anaphor.)

KLHM argue that this prediction that the root governing category is unmarked for the pronoun is wrong. The evidence they adduce is their claim that the root governing category (in which the pronoun is free from any c-commanding NP in the sentence) is not attested. I will argue (i) that their empirical claim is simply wrong, being based on a misconception of binding theory, and that (ii) even if the empirical claim were correct it would not necessarily be an argument against the Subset Principle.

Manzini and Wexler pointed out that pronouns tend to be associated with smaller governing categories, not with the largest (root) g.c., and provided some reasons for this. For reasons of space I will not go into these reasons. But KLHM argue that they know of no instances of the root governing category. They give examples of 10 languages in which they claim that the pronoun is free in the minimal governing category (like English).

But their linguistic evidence is quite weak. Consider, for example, one of the languages they mention, Japanese. The sole example[13] shows that a pronoun in Japanese may be coreferential with a NP outside of its minimal governing category. Their example is given in (4) (= KLHM (15)).

(4)   John-wa [Mary$_j$ — ni] [$_S$PRO$_j$ kare — ni]
        John$_i$ — TOP Mary$_j$ — DAT him$_i$ — DAT

        denwa — okake — sase — ta
        to — telephonecaused

But it is well known that the Japanese pronoun *kare* cannot be a bound variable.[14] An example (= (la) in Hoji (1990)) is given in (5):

(5)   daremo$_i$-ga[$_{NP}$[$_S$ zibun$_i$-ga/*kare$_i$-ga/ec$_i$ tukut-ta]
        everyone-NOM   self-NOM he-NOM   make-PAST

        omotya]-o kowasi-ta
        toy-ACC   break-PAST

        Everyone$_i$ broke the toy that he$_i$ had made.

In (5) *kare* must be free from *daremo*. That is, *kare* must be free from a quantifier which commands it, no matter how far up in the phrase-marker the quantifier is, that is in the root governing category. It may very well be that the example (4) doesn't represent a case of binding (of referential dependence in Higgenbotham's (1980) terms). Rather it may be a kind of "coreference". That is, in (4), *John* and *kare* may be contra-indexed, not co-indexed.[15] In fact, Hoji (1990) argues that *kare* is a kind of demonstrative and that it may be coreferential with, but not bound by other NP's. To get true evidence for binding one has to show that the pronoun may be a bound variable. The fact that the Japanese pronoun can't be a bound variable, no matter how far away the operator, provides evidence that the governing category for Japanese *kare is* the root sentence.

I haven't studied all the languages that KLHM mention, but of course one would expect that in many languages the pronoun can be a bound variable. At least one other language that KLHM mention in passing — Korean — also has a pronoun that is well-known to be incapable of being a bound variable. Thus the major empirical claim of KLHM is shown to be spurious.[16]

4. MARKEDNESS HIERARCHIES AND LEARNABILITY

The basic problem for linguistic theory is what Chomsky (1966), following Descartes, has called the problem of the "poverty of the stimulus," otherwise known as the problem of explanatory adequacy or learnability. The essential problem to be faced is how the child selects the correct grammar (parameter-setting in our case) given experience. It seems natural, on this view, for markedness hierarchies to play a role in solving the problem of explanatory adequacy or learnability. The Subset Principle does this, in a clear way; the problem of learnability is its primary motivation. The Stipulative Markedness Hierarchy (SMH), on the other hand, does not allow the problem of learnability to be solved. Thus it fails to meet the conditions of explanatory adequacy, the foundations of linguistic theory.

To see this, let us first consider the nature of the input (primary data) available to the learner. It has long been accepted that children do not receive negative evidence.[17] As shown by Manzini and Wexler (1987) and Wexler and Manzini (1987), a markedness hierarchy derived from The Subset Principle allows learning to take place under conditions of positive evidence only. How does the Stipulative Markedness Hierarchy fare on learning from positive evidence only?

For anaphors the Stipulative Markedness Hierarchy provides the same hierarchy as the Subset Principle does. Therefore, for anaphors, the learning properties of the SMH will be the same as those of the Subset Principle.

Now consider pronouns. The SMH stipulates that the markedness hierarchy for pronouns is the same as the hierarchy for anaphors. In particular, the English (smallest, i.e., domain of subject, see Manzini and Wexler (1987)) governing category for pronouns is the unmarked governing category, according to the SMH. But this creates a learnability problem, the very problem that the Subset Principle works perfectly on. If the governing category for a language L is syntactically larger (higher) than the English governing category, there will never be positive evidence to tell the learner to change the governing category. Thus the learner will always stay with the unmarked (i.e., small, English) governing category. There will be no way for the learner to move to the correct, higher, governing category. Thus the Stipulative Markedness Hierarchy fails on a fundamental theoretical and empirical criterion of linguistic theory, the problem of explanatory adequacy.

To take a concrete case, KLHM accept the existence of Manzini and Wexler's governing category c for pronouns, the minimal category which has a Tense, that is, the minimal finite clause. In particular, they accept Manzini and Wexler's characterization of Icelandic as having this value. Yet if a child starts out with the SMH's unmarked setting, whereby a pronoun must be free in the domain of a subject, there will never be positive evidence to tell her that, in fact, a pronoun must be free in the minimal finite clause, not in the domain of a subject. Thus the child will stay with the unmarked, incorrect, setting of the parameter, the pronoun having to be free in the domain of a subject. The final state of the language for this Icelandic child (adult) will accept as grammatical (i.e., not violating Principle B) sentences in which the pronoun is free in the domain of the closest subject, but not free in the minimal finite clause. This is the wrong grammar for Icelandic.

KLHM also argue in their Appendix that in Gothic, in fact earlier Germanic, pronouns were free in governing category c, the domain of the minimal finite clause. They summarize their evidence in the following way.

Thus, the Gothic pattern — reflexives but not pronouns with reference to the main clause subject — is directly evidenced in Gothic, Icelandic, and some dialects of Danish, and supported by indirect evidence in Middle High German and Old Norse, where proximate pronouns never or at most very rarely are found to occur in such constructions. Moreover, in the case of many of the languages exhibiting the English-type pattern (pronouns allowed with reference to the matrix subject), e.g. in German, Danish and Dano-Norwegian, there is evidence that this pattern is a result of a historical development.

Thus, apparently there is a fairly large group of languages which have the minimal finite clause as the governing category for pronouns. Manzini and Wexler (1987) pointed out, while hypothesizing that the governing category was parameterized for pronouns as well as anaphors, that there did not appear to have been as much study of the possibilities of variation for pronouns as for anaphors. But if KLHM are right there are indeed significant possibilities of variation for pronouns, as suggested by Manzini and Wexler.

Since the Stipulative Markedness Hierarchy fails on the essential grounds of explanatory adequacy (learnability) it is clear that it cannot be right as the markedness hierarchy for pronouns. One could attempt to stipulate the *opposite* hierarchy for pronouns, that is to stipulate one hierarchy for anaphors, one for pronouns, each of them agreeing with the hierarchies derived from the Subset Principle. Not only is this completely at variance with the spirit and arguments of KLHM, but it just reproduces the results of the Subset Principle.

There is a small section in KLHM in which they realize that the Stipulative Markedness Hierarchy as applied to the Icelandic and similar cases "would appear to leave a learnability problem (a 'Subset Problem') unsolved." They suggest the following possibility as a solution. Suppose that Manzini and Wexler's *Lexical Parameterization Hypothesis* doesn't hold, but that parameters are "initially fixed for grammars, not lexical items." Thus the learner of Icelandic will take the unmarked governing category a, the domain of subject, to be the g.c. for both anaphors and pronouns. When the learner hears a sentence in which an anaphor is bound in governing category c, the minimal finite clause, then the learner will change the governing category value to c, and this will apply to both anaphors and pronouns, despite the fact that the evidence was for anaphors only. This is because there is only one governing category parameter.

It is immediately clear that this proposal cannot begin to work, not even for the range of evidence alluded to by KLHM. The proposal clearly entails that languages have only one governing category, that applies to both anaphors and pronouns. But this is clearly false. It is false, for example, of the range of languages mentioned by KLHM in which the anaphor was completely long-distance (e.g. Chinese) (thus the anaphor has g.c. value e, the root value), but for which KLHM claim that the pronoun is free in the smallest (value a) governing category.[18] Manzini and Wexler gave evidence to show that anaphors and pronouns in the same language may have different governing categories, and KLHM accept this result. Therefore their suggestion, as they themselves seem to realize, cannot begin to be right.

It is worth pointing out, in addition, that the hypothesis that linguistic parameters are associated with grammars, as opposed to lexical items, or something close to lexical items, is seriously questioned in contemporary linguistic theory. Thus, in addition to Manzini and Wexler's *Lexical Parameterization Hypothesis* and Wexler and Chien's (1985) *Lexical Learning Hypothesis*, Borer (1984) has proposed that parameters are associated with morphological items. Fakui (to appear) and Chomsky (1986b) have strengthened these proposals to the hypothesis that parameters are associated with functional categories. So, the proposal that the governing category parameter is associated with grammars will very likely turn out to be not consistent with the results of linguistic theory.

Thus, the Stipulative Markedness Hierarchy is left without a possible solution to the fundamental problem that seems to arise in this empirical domain. Of course it is always possible that some proposal might make it compatible with the basic facts, but with anything like current understanding there does not seem to be a way to make it into a reasonable proposal. The markedness hierarchy deduced from the Subset Principle, however, solves the learnability problem, in exactly the right way.

5. LANGUAGE ACQUISITION

The markedness hierarchy deduced from the Subset Principle not only solves the learnability (explanatory adequacy) problem for pronouns and anaphors. It also provides a prediction as to the order of acquisition. That is, we would predict that a child hearing a language with a more marked value of a parameter might sometimes have a grammar with a less marked setting.[19] For example, the markedness hierarchy for anaphors would predict that in languages with long-distance reflexives, the small (domain of subject) governing category may be selected for a while by the child. Chien and Wexler (1987a) found this result for Chinese, Lee and Wexler (1987) found the same result for Korean and Hyams and Sigurjónsdóttir (1991) also found the result for Icelandic.[20]

KLHM accept this reasoning. They attempt to argue from well-known developmental data that the acquisition of governing categories for pronouns does not follow the predictions of the Subset Principle induced markedness hierarchy. Wexler and Chien (1985), Chien and Wexler (1987b, 1991), among many other studies have shown that children up to a fairly late age (e.g. about age 6;0) accept a pronoun with a local c-commanding antecedent, for example, they would accept a sentence like (6), where *her = Mary*.

(6)   Mary saw her

This is a very robust finding, having been replicated in a number of languages and with a large variety of methods.[21] KLHM argue that this finding shows that children have a too small governing category for the pronoun, even smaller than the correct English value (domain of subject, Manzini and Wexler's value *a*). Thus, they argue, not only don't children show a too large value, which the Subset Principle hierarchy predicts might happen, but they show a value smaller than any of those allowed by the hierarchy. Thus, they argue, Stipulative Markedness provides a better account of the language acquisition data, since the unmarked value for SM is the small governing category *a*, the domain of a subject.

This argument fails on two counts. First, note that if it were the case that children were violating Principle B when they accepted sentences like (6), then not only does the Subset Principle derived markedness hierarchy not explain this, but neither does Stipulative Markedness. SM says that the unmarked category is the English value, domain of subject. Such a governing category predicts that (6) is ungrammatical, by Principle B. Just because the unmarked value for SM is a smaller (more minimal) category than it is for the Subset Principle markedness hierarchy still doesn't make (6) grammatical for SM. Thus, if (6) represents a Principle B violation, it is as much a problem for Stipulative Markedness as it is for the Subset Principle.

Second, KLHM don't discuss much of the relevant developmental literature, both as to its empirical details and its conclusions. The conclusion of the work that they mention is that children *do* know Principle B (and the correct governing category) when they accept sentences like (6). Wexler and Chien (1985) first suggested that children do know Principle B despite their acceptance of sentences like (6), suggesting that the coreference was an accidental coreference, and that children would not allow a pronoun to be bound by a local c-commanding NP.[22] This prediction was confirmed in Wexler and Chien (1988) and Chien and Wexler

(1991), where it was shown that children who accept sentences like (6) still will not accept sentences like (7), where *every woman* is co-indexed with (binds) *her*, so that the sentence means that every woman saw herself.

(7)   every woman saw her

Wexler and Chien's conclusion is that children know Principle B, though they are missing some pragmatic knowledge. Confirming experimental results may be found in McDaniel, Cairns and Hsu (1991), Thornton (1990), Thornton and Wexler (1991) and Avrutin and Wexler (1991). A related interpretation of the results may be found in Grodzinsky and Reinhart (to appear). Grimshaw and Rosen present a slightly different interpretation of the results, but they agree on the essential idea that children know Principle B.

Thus the developmental literature argues quite strongly that children know Principle B; in fact, to my knowledge, there is no argument in the literature that children don't know Principle B, except for the suggestion by KLHM.[23] Sentences like (6), when they are accepted by children, just aren't binding theory violations. Thus they are not violations of a particular governing category. In short, not only do they not distinguish between the Subset Principle markedness hierarchy and Stipulative Markedness, but they are completely irrelevant to the question of whether children know the governing category.[24]

There is the question of whether children ever show that they have too large a governing category. As we have seen, the Principle B evidence that has been such a focus of research is irrelevant to that issue, since it shows only that children do have the correct governing category at a certain age. How about at earlier ages? There is really no study which directly deals with this question. And it would be somewhat difficult to tell. Since children at the youngest ages don't produce, and might not understand, complex sentences, probably for reasons having nothing to do with syntax, we really can't test the largest (root) governing category at the younger ages. Possibly we could test other values of the governing category. For example, is there a stage where children can't accept sentences like (8), where *her* is referentially dependent on *Mary*? Or even more clearly, do the children reject (8b)?

(8) a.   Mary likes her book
    b.   every woman likes her book

If so, then we could argue that the children believe that the pronoun must be free in the domain of tense.

Although we do not have information of this kind, one relevant fact *is* clear. As Wexler and Manzini (1987) point out, the earliest usages of pronouns in natural speech are in situations where they are free, that is, they are not co-indexed (or coreferential) with anything else in the sentence. Thus it is quite possible that at the earliest stages children believe the pronoun must be free. Stronger experimental tests are desirable. Nevertheless, it is quite conceivable, and quite consistent with the evidence to hypothesize that there is a stage in which children take pronouns to have the largest (root) governing category. If we did not find such a stage it would still be possible that children simply made instant use of evidence which shows that pronouns don't have to be free in any category higher than domain of subject in English.[25] The conclusion is clear: developmental evidence is quite consistent with the unmarked status of root governing category for pronouns.

6. THE MATHEMATICAL THEORY OF LEARNABILITY AND THE HISTORY OF THE SUBSET PRINCIPLE

Summarizing to this point, we see that the Subset Principle is an intensional principle; that its linguistic implications for binding theory seem empirically correct, at least with respect to the range of data considered by KLHM; that its developmental predictions have not been disconfirmed; that, in strong contrast to KLHM's Stipulative Markedness, it solves the problem of explanatory adequacy (learnability) for the binding domains and that, again in contrast to Stipulative Markedness, it fits into principles and parameters theory, deriving markedness hierarchies from the interaction of principles rather than from a stipulation (analogous to a rule). There is one other attempted class of arguments which KLHM allude to. These have to do with the relation of the Subset Principle to the mathematical theory of learnability.

First we should comment on the inaccurate and misleading account of the history of the Subset Principle in KLHM. They write that the Subset Principle "derives from studies of computational induction in a framework of recursive function theory (Angluin 1980)." This is false. Perhaps KLHM's misconception arises from a misreading of Berwick (1985). It is true that Berwick bases part of his discussion of The Subset Principle on a theorem in Angluin (1980). This is a representation theorem for learnability from positive evidence (similar to theorems in Wexler and Hamburger (1973), Wexler and Culicover (1980)). The name — Subset Principle — that is usually used in the linguistic and developmental linguistic literature is taken from this reference. But the concept itself that is used is *not* this Subset Principle, since the concept that is used in current theory is one that applies to the parameter-setting case, in which there are only finite possibilities. Thus, Manzini and Wexler adopted the name "Subset Principle" from Berwick, but provided a formal definition of the Subset Principle that was not the same as his.

Moreover, a large variety of algorithms are consistent with the theorem of Angluin (1980). The theorem only requires that an effective ordering of triggering sets of data exists. It does not specify how selection shall be made. And it is true that, even in the finite parameter-setting case, selection could be made in a large variety of ways. All sorts of bizarre algorithms can be formulated. For example, suppose that parameter-setting X yields the language L(X) with sentences a and b while parameter-setting Y yields the language L(Y) with sentences a, b and c. An algorithm A which learns from finite data could say the following: "When sentence a is encountered, but b has not been encountered, select the value Y. When sentence b is encountered, select the value X. When sentence c is encountered select the value Y." Algorithm A will ultimately select the correct language. It only has the "funny" property of selecting Y at a point where only examples of sentence a have been heard. This contradicts the (linguistically-based, e.g. Manzini and Wexler's) Subset Principle, but learning will take place (under the usual assumption of formal learning theory (cf. Gold 1967) that no data is systematically excluded). Manzini and Wexler's claim is that the Subset Principle is *sufficient* for learning. It solves the learnability problem. And it is a particularly natural, general, simple algorithm, which applies to a large number of cases in general, in contrast to the more complicated and specific algorithm A. This is just the kind of consideration that one uses in providing evidence for linguistic principles.

The Subset Principle and its usages in linguistic theory and the theory of language acquisition derive from attempts to solve the problem of explanatory adequacy (learnability) in the context of the assumption of no negative evidence. Wexler and Hamburger (1973) made the claim that the child wasn't exposed to negative evidence and argued that linguistic theory and learning theory had to accommodate this fact. Baker (1981) proposed that when a verb is such that it can take two sets of structures, one of which is included in the other, the human learner will select the smaller set first, so as to avoid the problem of no negative evidence. For example, a verb will be considered to not take a double object construction until positive evidence is received that it does. This "conservative" learning is an application of the logic of the Subset Principle. Williams (1981, based on his earlier work) proposes a model for the learning of specifiers and complements in X-bar theory which essentially uses the Subset Principle; the paper essentially proposes that the Subset Principle exists (not called that, of course). Berwick (1985, chapter 6) describes a large class of learning problems for linguistic structures, the solutions to which are applications of the Subset Principle, in fact the Subset Principle applied to the finite case (that is, the same case as later studied by Manzini and Wexler). Rizzi (1982) suggests, essentially on Subset Principle (no negative-evidence) considerations, that the unmarked setting of the null-subject parameter would have −null-subject (i.e., subjects obligatorily present). Faced with empirical discoveries that children start with null subjects,

Hyams (1985) takes pains to insure that the theory of syntax is such that the Subset Principle is not violated even if null subjects represent the unmarked state. There are a large number of other cases in the linguistic literature which are essentially applications of the logic of the Subset Principle.

Williams (1987) argues that "at the juncture of grammatical studies and acquisition studies," . . . there is "one research-generating idea". This is the "no-negative-evidence hypothesis."[26] Basically, as Williams points out, the Subset Principle is just the logic of the "no-negative-evidence hypothesis." The examples that Williams discusses, including his (1981) work on the setting of parameters in the base component, demonstrate again that the Subset Principle derives from attempts to solve the learnability puzzle for natural language.

Nowhere in KLHM is mention made of this history.

Perhaps this misreading of the history of the Subset Principle is what lies behind their frequent reference to the Subset Principle as used in linguistic theory and in language acquisition as the "informal Subset Principle." But even if the Subset Principle *did* derive from the mathematical theory of learnability, it would be inaccurate to call it the "informal" principle. There is nothing "informal" about the Subset Principle. It is precisely, even formally, even mathematically, defined (e.g. in Manzini and Wexler 1987) and there is no confusion as to its application. Nor do KLHM claim otherwise. But nevertheless they call it the "informal" principle.

KLHM write that in Kapur, Lust, Harbert and Martohardjono (1990) they "consider the question of the connection between this informal SP and its counterpart in formal learning theory, and conclude that it is neither equivalent to nor entailed by the original formal learning theory from which it is descended. We consider the relation of this proposal to a strong form of a theory of UG which does not consult a learnability module."[27]

I don't have the paper referred to available to me. However, the presentation that KLHM made at the conference on which this volume is based had their argument in it. Basically they claimed that the Subset Principle wasn't sufficient for learning, that even with the Subset Principle there were classes of languages[28] that were unlearnable. Thus they challenged the claim of Manzini and Wexler (1987) and Wexler and Manzini (1987) that the Subset Principle was sufficient for the setting of parameters.[29]

I will not go into their examples here, since I don't have them in written form, and they are not given in the paper from this volume. However an essential property of their examples was evident on inspection. The artificial examples which showed that the Subset Principle wasn't sufficient for learning contained parameters with infinite numbers of values. This is an impossibility in any linguistic theory under discussion today, that is, in any principles and parameters theory, e.g. Chomsky (1981). Moreover, an infinity of values of a parameter means that there are an infinity of possible grammars, again completely contradicting a fundamental result of principles and parameter theory, from which it is derived that there are only a finite number of possible grammars.[30] Moreover, it is completely clear that KLHM's examples *necessarily* depended on their being an infinity of possible grammars. If they made the accepted assumption that in natural language there are only a finite number of possible grammars, then it became clear that the Subset Principle was sufficient for learning.

Thus KLHM's "mathematical" arguments are not valid for natural language. In fact, it is clear, as pointed out to me by a participant in the conference, that if their arguments necessarily depend on an empirically wrong assumption, then the arguments in fact lend support to the Subset Principle, which is the principle the sufficiency of which depends on the empirically correct assumption.

7. SUMMARY AND CONCLUSION

We have concluded that the arguments for Stipulative Markedness and the arguments against the Subset Principle do not seem to hold much weight upon inspection. Foundational issues sometimes are difficult, especially when new proposals are being entertained. Thus we should expect some confusion.

There is more to the discussion, however, than simply the question of whether a particular principle is correct. Part of the reason for the different views involves the conception of markedness and the goals of linguistic theory. We have followed the view that the central problem of linguistic theory is the problem of learnability (of explanatory adequacy, in Chomsky's terms). We have related markedness hierarchies to the solution of this problem.

The Stipulative Markedness Hierarchy, on the other hand, is meant to be relevant to other kinds of considerations. To see this, consider KLHM's conclusion that "In any case, pronouns observing GC(e) [the root governing category] are at least infrequent if they exist at all . . .". KLHM's argument reduces to the statistical infrequency of the root governing category for pronouns. In my view this is the heart of the matter. What SMH is designed to predict is the frequency of particular parameter-settings across known languages. This is the traditional view of markedness in structural linguistics. For example, this is the view of markedness exemplified by Greenberg (1966). Greenberg believes that universals are statistical tendencies and that it is the task of linguistics to explain statistical tendencies. KLHM seem to share this view.

The view we have argued for in this paper is that markedness hier-archies (and the Subset Principle) are relevant to the central problem of the poverty of the stimulus, in fact, are part of the solution to that problem. We have shown how markedness hierarchies can be made to follow in a principled fashion and solve the learnability problems at the same time. Direct empirical evidence, both in terms of adult grammars and developmental phenomena, also supports the Subset Principle and this view of markedness. Moreover, we have shown that the Subset Principle, just as other linguistic principles, is a part of the I-language.

In summary, we have seen that the Subset Principle fares rather well on such basis issues. Of course, the problem of the setting of parameters is a complex problem, and it would be very surprising if the Subset Principle were the only principle needed, in addition to the fitting of the parameters to evidence. Probably there should be more structure to the problem. This structure will depend on the creation of more complex theories of parameters, possibly with hierarchical properties, etc. Moreover, much of the development of language may be maturational.[31] There are large numbers of such issues, and investigation of the exact role of the Subset Principle will involve working out of much of this. But the Subset Principle seems to be quite central to the discussion of how experience sets properties of linguistic variation, such as parameters. That is, the Subset Principle seems relevant (along with a large number of other principles and properties) to the solution of Plato's Problem.

Whether binding theory variation is to be explained in terms of the kind of parameterization in Manzini and Wexler is a different question. As pointed out earlier, there are a number of approaches to variation in binding which might suggest a different approach in at least some cases. For example, movement of anaphors at LF might account for some of the variation. But the variation will still have to be stated someplace in the grammar, and there is reason to believe that the Subset Principle might turn out to be relevant to those cases. At any rate, it is clear that we are only in the beginning phases of the attempt to understand how learning of the variational aspects of language takes place, but we have already made enough progress to encourage us to continue in the search for the underpinnings of this learning.

NOTES

* I would like to thank Eric Reuland for a large amount of help with this paper. Thanks also to an anonymous reviewer. This work was supported by NSF grants BNS-8820585 and DBS-9121201.
[1] In Wexler (1990) I called this mechanism the "General Motor of Learning."
[2] This is in the simplest case, with no further structure imposed on the problem. It is sometimes suggested that "indirect" negative evidence may be used (Chomsky 1981). For arguments and discussion, see Wexler (1987).
[3] Of course there have been other approaches to a deductive structure for markedness, for example, Chomsky's (1955) concept of *evaluation metric*, which was proposed in conjunction with a hypothesis-testing model of learnability. The Subset Principle, however, seems

particularly natural for a system in which the learning decisions are much smaller, in particular for parameter-setting. See the discussion that follows.

[4] Actually, it is something of a misnomer to call the Subset Principle a "new concept." As we will discuss, the Subset Principle has actually been assumed, often implicitly, by a good deal of linguistic theory.

[5] See Chomsky (1965) and many other publications.

[6] For a discussion of the role of the argument from the poverty of the stimulus in linguistic theory and in cognitive science more generally, see Wexler (1991).

[7] The Subset Principle can be extended to apply to other kinds of variation (e.g. lexical properties), not only the setting of parameters, but for simplicity, for the remainder of this paper I will only consider the case of parameter-setting.

[8] E.g., Manzini and Wexler (1987) intended the Subset Principle as an I-language principle. Consider, for example, their proofs that one value of the governing category parameter was unmarked with respect to another value. These were carried out by showing that any derivation which violated the marked value also violated the unmarked value.

[9] I am being informal with respect to the definitions of the governing categories. Precise definitions may be found in Manzini and Wexler (1987).

[10] For simplicity I have been giving definitions only for cases in which there is one parameter. A complete account would have to consider the multi-parameter case. It seems clear that the definitions will be generalizable to the multi-parameter case in a way analogous to the definitions in Manzini and Wexler (1987) and Wexler and Mazini (1987). The Independence Principle would have to be reformulated in the terms given here, but all this seems straight-forward.

[11] Alternative analyses in which, for example, anaphors move at LF, are not at issue here. See, among others, Lebeaux 1983, Pica 1987, Battistella 1987, Huang and Tang 1989, Reinhart and Reuland 1991, Wexler, in press.

[12] Note that there is no special status to "syntactically minimal domain" except to the extent it participates in particular syntactic modules. Principle C, for example, says that a NP is free in the entire sentence. This does not mean that there is a special status to "syntactically maximal domain."

[13] They give only one example for each language.

[14] See Saito and Hoji (1983), Hoji (1990) and references cited there.

[15] See Chien and Wexler (1991) for an application of this well-known idea in linguistic theory to the development of binding theory.

[16] Manzini and Wexler suggested 2 cases in which the pronoun was free in the root governing category. The first case is a pronoun in Yoruba which has to be free (from subjects only) in the root governing category. KLHM point out that Pulleyblank (1985) has suggested an alternative analysis. The second case are epithets such as 'the idiot', which KLHM suggest can be treated as names, subject to Principle C. Since not much analysis is presented, I will not discuss these cases here. Even if they turned out to be correct, they would not violate the Subset Principle, only take away some of the cases adduced for it. KLHM (note 6) also point out a possible other case of root governing category of a pronoun, from Tamil. Again, they suggest there may be another analysis. But, as in the other cases, they do not provide a convincing alternative analysis.

[17] Wexler and Hamburger (1973) first proposed that children do not receive negative evidence and that this provides a problem for theory to solve. Their proposal was based on empirical results, especially Brown and Hanlon (1970). Baker (1979), Wexler and Culicover (1980) and many other papers have accepted the "no negative evidence" assumption. For some alternatives, e.g. the breaking of "Concreteness", see Wexler (1987).

[18] Although they do not say so, perhaps this is the reason that KLHM seem to have serious doubts about the proposal. They write, "we do not suggest this scenario as a veridical proposal . . ."

[19] We wouldn't always, or even often, expect to find this phenomenon in the developmental data we look at, because the child might have passed through this stage very quickly, even instantly, if the relevant input were available. The crucial prediction is that we never find the opposite state, where the child selects a more marked setting for the parameter than the grammar which she is learning has.

[20] The effect lasts to an old enough age that it may very well be *too* strong; on a strict Subset Principle view, why doesn't the input have an effect before it does? See Wexler (in press) for an alternative proposal.

[21] See Chien and Wexler (1991) for a listing of much of the literature.

[22] See also Montalbetti and Wexler (1985).

[23] Jakubowicz (1984) suggests that children miscategorize the pronoun as an anaphor, and that is why they accept (6). As Wexler and Manzini (1987) point out, if Jakubowicz intends that the children only treat pronouns as anaphors, this cannot be correct, because the earliest (before age 2 usually) use of a pronoun is as a completely (syntactically) free form. An anaphor cannot be free; therefore the children can't be treating the pronoun as an anaphor when they use it freely. If, one the other hand, the suggestion is that the pronoun is a homophonous form for the child, both a pronoun and an anaphor, then why do the children reject forms like (7), which would be good if the pronoun were an anaphor? So miscategorization cannot account for the empirical data. See Avrutin and Wexler (1991) for discussion. It should be pointed out that it has been accepted for a long time in the developmental field, based on data such as this, that the children are not treating the pronoun as an anaphor; as in linguistics, the patterns of empirical data can be telling on this point.

[24] Based on the limited range of data that KLHM discuss, there is a suggestion that they might make that they don't. They might suggest that there is an additional value of governing category, say VP, and that the children who accept (6) have chosen that governing category. Koster (1987) has suggested that there are cases in Dutch dialects where there is a smaller governing category than domain of subject. Varela (1989) has suggested that children who violate (6) have chosen VP as the governing category. KLHM could suggest that VP is the unmarked governing category and that is why children have chosen it. There would be major problems with this proposal, however. First, as McDaniel and Maxwell (1991) have shown experimentally, Varela's proposals do not work for children. Namely, the anaphor and pronoun domains cannot simultaneously be VP. Second, the same learnability problem as we discussed earlier would hold; how would the children change their governing category. Third, the proposal cannot account for children's rejection of (7), for their knowledge that a pronoun cannot be locally bound. A large number of empirical and theoretical arguments converge on the conclusion that children at the age under consideration know Principle B and the correct governing category.

[25] KLHM also discuss some data from Padilla-Rivera (1985, 1990) which shows that in some cases children prefer a sentence internal antecedent to an external one, when both are grammatical. Thus in sentence (i) from Padilla-Rivera (= KLHM's (27b)), the pronoun *ella* is taken more often to be *la pantera* than it is taken to be an external referent.

(i)   La pantera patea el avioncito detras de ella
         The panther kicks the plane-(dim) behind her

Actually, this is a well-known finding, especially in pragmatic contexts where the external referent isn't highlighted. But it is not at all surprising. From the standpoint of governing categories, all it means is that children by this age (3;0—3;11 is the youngest age range tested) have already learned that pronouns don't have to be free in a higher governing category. The evidence will be abundant. An even stronger experimental result to show that children have a small governing category at this age would be if the pronoun were clearly bound, e.g. by a quantifier or by a *wh*-phrase. At any rate, such data doesn't constitute evidence that the unmarked governing category isn't the largest (root) one.

[26] For some recent work which attempts to go beyond no negative evidence considerations in the theory of parameter-setting, see Gibson and Wexler (1991). Also see Clark (to appear).

[27] This last sentence is clearly a non-sequitur. From an empirical point of view, there can be no such thing as a theory of UG which does not consult a learnability module. Since exactly which grammar is chosen depends on experience (if only for parameter-settings, or lexical values, or whatever minimal amount of learning occurs), there must be a learning component. As I have pointed out, the minimal learnability component involves fitting the parameter-settings to input (the General Motor of Learning, Wexler 1990). Every theory in generative grammar assumes at least this much of a learnability component.

[28] In their examples, KLHM used the notion of E-language, despite their statements in this paper that one shouldn't use such notions.

[29] Of course, as discussed in Manzini and Wexler, in cases where the Subset Principle doesn't apply, learning takes place simply by fitting the parameter to the evidence. This fitting of learning to evidence is implicit in the definition of the Subset Principle. The claim that the Subset Principle is sufficient (that is, the theorem that the Subset Principle is sufficient) means that in addition to fitting parameters to data, only the Subset Principle is needed, at least for the kinds of parameters discussed in Manzini and Wexler. KLHM give no further kinds of parameters, so that the theorem applies to any case they discuss as well. See Gibson and Wexler (1991) for arguments that, given the idea of parameter-setting that seems to underlie Principles and Parameters theory, there are linguistically valid parametric systems that can't be set simply by fitting the parameter to the data. Further structure (beyond the Subset Principle) must be given to the learner.

[30] See Chomsky (1981). And for 2 different attempts to derive this property of "strong nativism" from learnability considerations see Osherson and Weinstein (1982) and Wexler (1981).

[31] See Borer and Wexler (1987).

REFERENCES

Angluin, Dana: 1980, 'Inductive Inference of Formal Languages from Positive Data', *Information and Control* **45**, 117–135.

Baker, C. Lee: 1979, 'Syntactic Theory and the Projection Problem', *Linguistic Inquiry* **10**, 533–581.

Battistella, Edward: 1987, 'Chinese Reflexivization', paper presented at the 2nd Harbin Conference on Generative Grammar, Heilongjiang University, Harbin, People's Republic of China, manuscript, University of Alabama at Birmingham.

Battistella, Edward: 1989, 'Chinese Reflexivization: A Movement to INFL Approach', *Linguistics* **27**, 987–1012.

Berwick, Robert: 1982, *Locality Principles and the Acquisition of Syntactic Knowledge*, Doctoral dissertation, Department of Electrical Engineering and Computer Science, MIT.

Berwick, Robert: 1985, *The Acquisition of Syntactic Knowledge*, MIT Press, Cambridge, Massachusetts.

Borer, Hagit and Kenneth Wexler: 1987, 'The Maturation of Syntax', in T. Roeper and E. Williams (eds.), *Parameter Setting*, Reidel, Dordrecht, 123–172.

Brown, Roger and Camille Hanlon: 1970, 'Derivational Complexity and the Order of Acquisition of Child Speech', in J. R. Hayes (ed.), *Cognition and the Development of Language*, Wiley, New York, 11–53.

Chien, Yu-Chin and Kenneth Wexler: 1987, 'Comparison Between Chinese-speaking and English-speaking Children's Acquisition of Reflexives and Pronouns', presented at the 12th Annual Boston University Conference on Language Development.
Chien, Yu-Chin and Kenneth Wexler: 1991, 'Children's Knowledge of Locality Conditions in Binding as Evidence for the Modularity of Syntax and Pragmatics', *Language Acquisition* **1**, 225—295.
Chomsky, Noam: 1965, *Aspects of the Theory of Syntax*, MIT Press, Cambridge, Massachusetts.
Chomsky, Noam: 1966, *Cartesian Linguistics: A Chapter in the History of Rationalist Thought*, Harper & Row, New York.
Chomsky, Noam: 1975, *Logical Structure of Linguistic Theory*, Plenum, New York; drawn from an unpublished 1955—56 manuscript.
Chomsky, Noam: 1981, *Lectures on Government and Binding*, Foris, Dordrecht.
Chomsky, Noam: 1986a, *Knowledge of Language: Its Nature, Origin, and Use*, Praeger, New York.
Chomsky, Noam: 1986b, *Barriers*, MIT Press, Cambridge, Massachusetts.
Clark, Robin: 1992, 'The Selection of Syntactic Knowledge', *Language Acquisition* **2**.
Fukui, Naoki: to appear, 'Deriving the Differences Between English and Japanese: A Case Study in Parametric Syntax', *English Linguistics*.
Gibson, Edward and Kenneth Wexler: 1992, 'Parameter Setting, Triggers and V2', Presented at the 15th GLOW Colloquium, Lisbon, Portugal.
Gold, E. Mark: 1967, 'Language Identification in the Limit', *Information and Control* **10**, 447.
Greenberg, Joseph: 1966, *Language Universals, with Special Reference to Feature Hierarchies*, Mouton, The Hague.
Grimshaw, Jane and Sarah Rosen: 1990, 'Knowledge and Obedience: The Developmental Status of the Binding Theory', *Linguistic Inquiry* **21**, 187—222.
Grodzinsky, Yosi and Tanya Reinhart: to appear, 'The Innateness of Binding and the Development of Coreference: A Reply to Grimshaw and Rosen', *Linguistic Inquiry*.
Higginbotham, James: 1980, 'Anaphora and GB: Some Preliminary Remarks', in J. Jensen (ed.), *Proceedings of the 10th Annual Meeting of Northeastern Linguistic Society* (NELS), University of Ottawa, Ontario.
Hoji, Hajime: 1991, 'Kare', in R. Ishihara and C. Georgopoulos (eds.), *Interdisciplinary Approaches to Language: In Honor of Professor S.-Y. Kuroda*, Kluwer Academic Publishers, Dordrecht, 287—304.
Huang, C.-T. James and C.-Z. Jane Tang: 1989, 'On the Local Nature of the Long-Distance Reflexives in Chinese', in *Proceedings of the North Eastern Linguistics Society 19*, GLSA University of Massachusetts, Massachusetts.
Hyams, Nina: 1986, *Language Acquisition and the Theory of Parameters*, Reidel, Dordrecht.
Hyams, Nina, and Sigridur Sigurjónsdóttir (1991)', Parametrizing Binding Theory: Evidence from Language Acquisition', manuscript, UCLA.
Kapur, Shyam, Barbara Lust, Wayne Harbert and Gita Martohardjono: 1990, 'On Relating Universal Grammar and Learnability Theory: Intensional and Extensional Principles in the Representation and Acquisition of Binding Domains', in preparation.
Kapur, Shyam, Barbara Lust, Wayne Harbert and Gita Martohardjono: this volume, 'Universal Grammar and Learnability Theory: The Case of Binding Domains and the "Subset Principle"'.
Koster, Jan: 1987, *Domains and Dynasties: The Radical Autonomy of Syntax*, Foris, Dordrecht.
Lebeaux, David: 1983, 'A Distributional Difference between Reciprocals and Reflexives', *Linguistic Inquiry* **14**, 723—730.

Lee, Heoin-Jin and Kenneth Wexler: 1987, 'The Acquisition of Reflexives and Pronouns in Korean', paper presented at the 12th Annual Boston University Conference on Language Development.

Manzini, Rita and Kenneth Wexler: 1987, 'Parameters, Binding Theory, and Learnability', *Linguistic Inquiry* **18**, 413—444.

McDaniel Dana and Tom Maxfield: 1991, 'Principle B and Contrastive Stress', manuscript, University of Southern Maine Gorham, Maine and University of Massachusetts, Amherst, Massachusetts.

McDaniel, Dana, Helen Cairns and Jennifer Hsu: 1991, 'Binding Principles in the Grammars of Young Children', *Language Acquisition* **1**, 121—138.

Montalbetti, Mario and Kenneth Wexler: 1985, 'Binding Is Linking', in the *Proceedings of the West Coast Conference on Formal Linguistics*, *4*, UCLA, Los Angeles, California.

Osherson, Dan and Scott Weinstein: 1982, 'A Note on Formal Learning Theory', *Cognition* **11**, 77—88.

Osherson, Dan, Michael Stob and Scott Weinstein: 1986, *Systems That Learn: An Introduction to Learning Theory for Cognitive and Computer Scientists*, MIT Press, Cambridge, Massachusetts.

Padilla-Rivera, Jose: 1985, *On the Definition of Binding in Spanish: The Roles of the Binding Theory Module and the Lexicon* Doctoral dissertation, Cornell University, New York.

Padilla-Rivera, Jose: 1990, *On the Definition of Binding Domains in Spanish: Evidence from Child Language*, Kluwer Academic Publishers, Dordrecht.

Pica, Pierre: 1987, 'On the Nature of the Reflexivization Cycle', *Proceedings of NELS 17*, GLSA, University of Massachusetts-Amherst, 483—499.

Pulleybank, Douglas: 1988, 'Clitics in Yoruba', in H. Borer (ed.), *The Syntax of Pronominal Clitics*, Academic Press, Orlando.

Reinhart, Tanya and Eric Reuland: 1991, 'Anaphors and Logophors: An Argument Structure Perspective', in Jan Koster and Eric Reuland (eds.), *Long-Distance Anaphora,* Cambridge University Press, Cambridge, England.

Rizzi, Luigi: 1982 *Issues in Italian Syntax*, Foris, Dordrecht.

Safir, Kenneth: 1987, 'Comments on Wexler and Manzini', in T. Roeper and E. Williams (eds.), *Parameter Setting*, Reidel, Dordrecht.

Saito, Mamuru and Hajime Hoji: 1983, 'Weak Crossover and Move Alpha in Japanese', *Natural Language and Linguistic Theory* **1**, 245—259.

Stowell, Tim: 1981, *Origins of Phrase Structure*, Doctoral dissertation, MIT, Cambridge, Massachusetts.

Thornton, Rosalind: 1990, *Adventures in Long-Distance Moving*, Doctoral dissertation, University of Connecticut, Storrs, Connecticut.

Thorton, Rosalind and Kenneth Wexler: 1991, 'Children's Acquisition of Bound-Variable Pronouns', to be presented at the 16th Annual Boston University Conference on Language Development.

Varela, Ana: 1989, 'A Structural Explanation of Children's Apparent Failure to Respect Condition B', paper presented at the 14th Annual Boston University Conference on Language Development.

Wexler, Kennth: 1982, 'On Extensional Learnability', *Cognition* **11**, 89—95.

Wexler, Kenneth: 1987, 'On the Nonconcrete Relation Between Evidence and Acquired Language', in B. Lust (ed.), *Studies in the Acquisition of Anaphora, Volume II: Applying the Constraints*, Reidel, Dordrecht, 33—42.

Wexler, Kenneth: 1990, 'On Unparsable Input in Language Acquisition', in L. Frazier and J. de Villiers (eds.), *Language Processing and Language Acquisition*, Kluwer Academic Publishers, Dordrecht, 105—117.

Wexler, Kenneth: 1991, 'On the Argument from the Poverty of the Stimulus', in A. Kasher (ed.), *The Chomskyan Turn*, Basil Blackwell, Cambridge, Massachusetts, 253—270.

Wexler, Kenneth: in press, 'Some Issues in the Growth of Control', in R. Larson, S. Iatridou, U. Lahiri and J. Higginbotham (eds.), *Grammar and Control*, Kluwer Academic Publishers, Dordrecht.

Wexler, Kenneth and Yu-Chin Chien: 1988, 'The Acquisition of Locality Principles in Binding Theory', paper presented at the 11th GLOW Colloquium, Hungary, March, pp. 28—30.

Wexler, Kenneth and Peter Culicover: 1980, *Formal Principles of Language Acquisition*, MIT Press, Cambridge, Massachusetts.

Wexler, Kenneth and Henry Hamburger: 1973, 'On the Insufficiency of Surface Data for the Learning of Transformational Language', in K. J. Hintikka, M. E. Moravcsik and P. Suppes (eds.), *Approaches to Natural Language*, Reidel, Dordrecht, 153—166.

Wexler, Kenneth and Rita Manzini: 1987, 'Parameters and Learnability in Binding Theory', in T. Roeper and E. Williams (eds.), *Parameter Setting*, Reidel, Dordrecht, 41—76.

Williams, Edwin: 1981, 'Language Acquisition, Markedness, and Phrase Structure', in S. Tavakolian (ed.), *Language Acquisition and Linguistic Theory*, MIT Press, Cambridge, Massachusetts, 8—34.

Williams, Edwin: 1987, 'Introduction', in T. Roeper and E. Williams (eds.), *Parameter Setting*, Reidel, Dordrecht, vii—xix.

WILLEM J. M. LEVELT

LEXICAL ACCESS IN SPEECH PRODUCTION

INTRODUCTION

Lexical access in speech production proceeds at a rate of, on the average, two to three words per second. At this rate words are selected from a production lexicon which contains thousands, and probably tens of thousands, of words. These words are not only selected, but also phonologically encoded. This happens at a rate of about 15 speech sounds per second. The problem to be addressed in this chapter is how these high-rate and fairly accurate processes of lexical selection and phonological encoding are organized.

Figure 1 outlines a possible architecture for the organization of these processes of lexical access. There is a so-called "formulator" receiving as input the (lexical) concept-to-be-expressed (usually as part of a larger

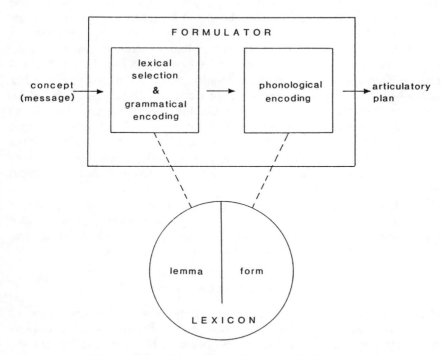

Fig. 1. Outline of a possible architecture for lexical access.

241

Eric Reuland and Werner Abraham (eds), Knowledge and Language, Volume I, From Orwell's Problem to Plato's Problem: 241–251.
© 1993 *by Kluwer Academic Publishers. Printed in the Netherlands.*

conceptualization) and producing as output an articulatory plan for the item (usually as part of a plan for a larger utterance). The formulator contains two component processors. The first one takes care of selecting the appropriate lexical item from the mental lexicon and of integrating it in the developing syntactic structure (grammatical encoding). The second one generates an articulatory program for the selected lexical item on the basis of its stored phonological code and the developing phonological context of the utterance as a whole.

Each of these two component processes may occasionally derail. If lexical selection goes awry, then errors such as these may occur:

Errors of lexical selection and grammatical encoding.
— Don't burn your toes (intended: fingers).
— Examine the horse of the eyes (intended: the eyes of the horse).

Failing phonological encoding leads to a very different kind of error:

Errors of phonological encoding.
— Fart very hide (intended: fight very hard)
— Face spood (intended: space food)

There are, in particular, three major issues to be asked with respect to the architecture in Figure 1. Two of these will be addressed in the present paper. The first one is the organization of lexical selection. What kind of architecture can mediate between conceptual structure (intentions, concepts to be expressed) and the knowledge base which is called the "mental lexicon", the speaker's store of information about the words of his or her language? The second issue is the structure of phonological encoding. Once an item has been retrieved from the mental lexicon, how is an articulatory program constructed out of the phonological information that is stored with the item? The third issue concerns the temporal alignment of lexical selection and phonological encoding. Are these two processes strictly successive in time, or can an item's phonological encoding begin **before** it is definitely selected?

Psycholinguistic research has largely concentrated on the second of these three issues, phonological encoding. I will, however, ignore it here (see Levelt (1989) for an extensive review of phonological encoding), and concentrate on lexical selection and on the time course issue.

LEXICAL SELECTION

How does the speaker select the appropriate word for the expression of the concept at hand? There are some rather heterogeneous proposals about the organization of this process (such as Morton's logogen theory, Goldman's discrimination net theory, Miller and Johnson-Laird's decision table theory, and the various connectionist suggestions, all reviewed in Levelt (1989)). What they have in common, however, is that they fail to solve what I have called the **hyperonym problem**. It can be formulated as follows:

The hyperonym problem. When item B's meaning entails item A's meaning, A is a hypernym of B. If B's semantic conditions are met, then A's are necessarily also satisfied. Hence, if B is the appropriate lexical item, A will (also) be retrieved.

So, for instance, **animal** is a hyperonym of **bear**. If the speaker intends to express the concept BEAR, then the semantic conditions for **bear** are satisfied, but — by implication — also those for **animal**. Hence **animal** will be retrieved (as well). All presently existing published theories fall in this hyperonym trap.

In order to solve this problem I have proposed that theories should implement two principles, the **specificity principle** and the **core principle**:

The specificity principle. Of all lexical items whose semantic conditions are satisfied by the concept to be expressed, the most specific one (the one with the largest potential of entailment) is to be selected.

The core principle. Each lexical item has a unique (set of) semantic core condition(s). An item is only retrieved if its core is satisfied by the concept to be expressed.

The specificity principle prevents that a target's hyperonym is selected (e.g. **animal** instead of **dog**). The core principle prevents that a target's subordinate is selected (e.g. **dog** instead of **animal**).

Ardie Roelofs, Manfred Bierwisch and I are presently developing an implementation of these principles in a logical network (see Figure 2).

Figure 2 shows a level of conceptual components or predicates, a level of semantic decision making, and a level of lexical items (more precisely: lemmas, i.e., lexical items without phonological specification). At the level of semantic decision making semantic functions are computed, given the input from the conceptual components. If a function evaluates to "true", activation is propagated to the lexical item on the next level. Such a function has, among others, two crucial properties. The first one is that its evaluation to true is only possible if its core component(s) is (are) satisfied. Second, the function cannot evaluate to true if it receives input from a hyperonym's core component. This "logical structure" of the network is complemented by an "activation structure" which handles the flow of information over time from the conceptual level to the lexical level. In order for an item to be selected, it is not only necessary that its test

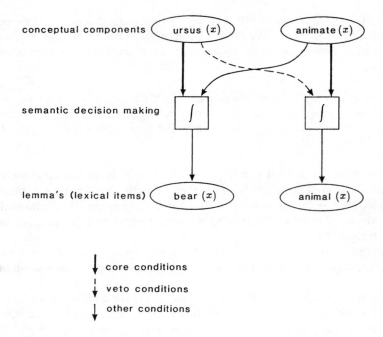

Fig. 2. A fraction of the logical net for lexical selection. The bold arrows depict the effects of core conditions (conditiones sine quibus non), the dotted arrows depict the working of veto conditions.

yields a positive outcome, but also that it attains a certain level of activation. In the present paper, I will not discuss this activation structure any further.

THE TIME COURSE OF LEXICAL ACCESS

Returning to Figure 1, one can distinguish two views on the time course of lexical access. The first one is the more traditional modular view, which says that there is no phonological encoding before lexical selection and there is, accordingly, no feedback from phonological encoding to lexical selection. The second view is the connectionist picture, which assumes a temporal overlap of lexical selection and phonological encoding, and a continuing interaction between the two processes. These two views are depicted in more detail in Figure 3.

In the classical theories (in particular Garrett's, Kempen's, Butterworth's, Levelt's — see Levelt (1989) for a review) there is an early phase of semantic activation, which rounds up in lexical selection. It is followed by a phase of phonological encoding where only the selected item becomes phonologically encoded. In the connectionist theories (in par-

THEORIES OF LEXICAL ACCESS

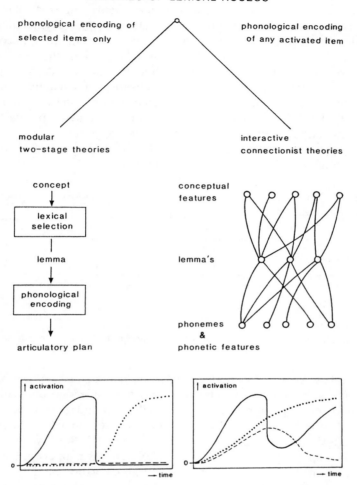

Fig. 3. Two theories of lexical access. The modular two-stage theory (top left) allows for the phonological encoding of the selected target item only. The connectionist theory (top right) predicts phonological activation of semantic alternatives to the target item. The corresponding activation curves are presented at the bottom: Semantic activation of target (——), phonological activation of target (. . . . .), and phonological activation of semantic alternative to target (---).

ticular Dell's, Stemberger's — see Levelt 1989), not only **selected** items are phonologically activated, but any semantically **activated** item.

There are three critical time course predictions proceeding from these theories; they are given at the bottom of Figure 3. The first one concerns the course of semantic activation. The modular theory predicts early, no late semantic activation; the connectionist theory (in particular Dell's) predicts early semantic activation and a late rebound of semantic activation, due to feedback from the phonemic to the lemma level. Second, the modular theory predicts late phonological activation, the connectionist theory predicts both early and late phonological activation. Third, the modular theory interdicts phonological activation of semantic alternatives (only the selected item, but no co-activated item becomes phonologically encoded). The connectionist theories, on the other hand, predict phonological activation of semantic alternatives to the target item.

We[1] performed several experiments to sort these predictions out. The experimental paradigm is presented in Figure 4.

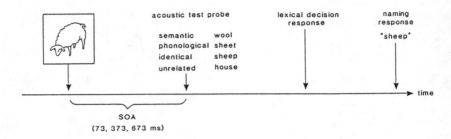

Fig. 4. The experimental procedure of the naming-cum-lexical decision task. See text.

Subjects were asked to perform a picture naming task. A long series of pictures was presented, one by one, and the subject would name each picture as soon as it appeared. Occasionally a secondary so-called "lexical decision" task was given. Shortly after presentation of the picture an acoustic test probe was presented, which could either be a word or a non-word, like **sip** (word) or **sef** (non-word). When this happened, the subject was supposed to push a "yes" or a "no" button, correspondingly. This task made it possible to probe into the subject's developing representation in his effort to produce the picture's name. For example, if the picture was one of a sheep the subject would internally generate semantic and phonological representations that were appropriate to the target name **sheep**. In order to test semantic activation of **sheep**, we would present as lexical decision probe a word like **wool**. There is reason to expect that semantic activation of **sheep** will delay the lexical decision to the acoustic test probe **wool**. Similarly, we could measure the phonological activation of **sheep** by presenting a test probe like **sheet**. In addition, the experiment contained the target word itself as probe (**sheep** in the example) — which I will not further discuss — and a control condition, namely a test word that is unrelated to the target, for instance **house**.

The critical issue is whether a semantic or a phonological test probe shows longer lexical decision latencies than the unrelated test probe. In order to see how the semantic and phonological representations develop over time, we presented the acoustic probe at different delays after the picture. There were three moments: 73, 373, and 673 ms. (on average) after picture onset. These are called "stimulus onset asynchronies", or SOAs.

The lexical decision latencies that we obtained in the experiment (192 subjects) are presented in Figure 5 (solid lines). In fact, these data are differences between the measured lexical decision latencies and lexical decision latencies for the same items when presented without concurring naming task. (The positive difference values in the figure show that the concurring naming task generally slowed down the lexical decision response.)

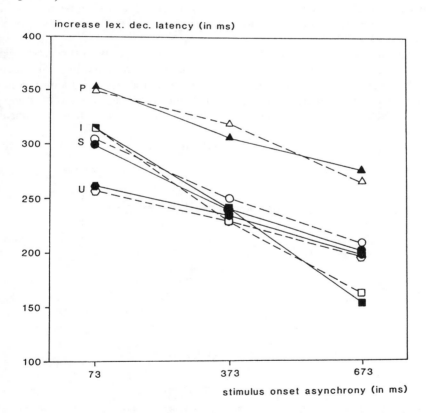

Fig. 5. Increase of lexical decision latencies in the dual naming/lexical decision task (solid lines) for three different types of probe and three SOAs. S = semantic probes, P = phonological probes, I = identical probes, U = unrelated probes. The dotted lines show the fit of the two-stage model to the data.

As both kinds of model predict, there is good evidence for early semantic activation (at the 73 ms. SOA the latency for the semantic probe is significantly longer than the latency for the unrelated probe). There is, however, no late semantic activation — contrary to the connectionist prediction. As both models predict, there is good evidence for late phonological activation, but seemingly contrary to the modular two-stage model there is evidence for early phonological activation. Hence, these results seem to be equivocal.

I will now argue that the two-stage model should be preferred. The argument is two-way. First, I will show that the two-stage model can give a perfect account of these data. Second, I will report experimental results on the phonological activation of semantic alternatives, which are in support of the two-stage model.

The data in Figure 5 are the statistical result of a huge number of measurements. It is therefore necessary to make a statistical model of this naming-cum-lexical decision task. Figure 6 depicts the model we[2] developed.

This model incorporates the two-stage modular theory in that there is a strict succession of lexical selection and phonological encoding. The idea is that there will be interference when the semantic stage of naming coincides with the semantic stage of lexical decision, and when the phonological stage of naming coincides with the phonological stage of lexical

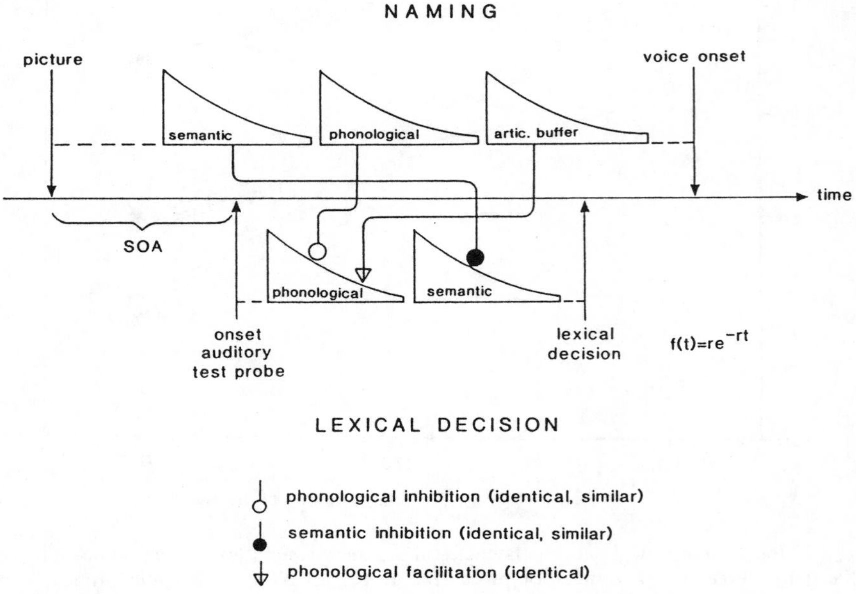

Fig. 6. Mathematical rendering of the two-stage model.

decision, in case same or similar items are involved in naming and lexical decision. The statistical time distribution of each of these phases is assumed to the exponential, with a characteristic rate parameter for each of the component processes. These rate parameters and the interference parameters can be estimated in order to find a best fit of the model to the data. This we did, and the result is presented in Figure 5, dotted lines. It shows that the data do not contradict the two-stage model. In fact, the fit is statistically perfect.

Turning now to the issue of phonological activation of semantic alternatives, I will report on an experiment that is quite similar to the previous one. But there are two differences. First, we used the short SOA (73 ms.) only, because it gave us both good semantic and good phonological activation in the previous experiment. Second, we used new acoustic test probes. Using again the example where the picture shows a sheep, we used the acoustic probe **goat** as a semantic probe, and the word **goal** as a phonological probe. This means that we can test whether the semantic alternative **goat** is not only semantically, but also phonologically active. In the latter case we should find an effect on **goal**. And that is what the connectionist theories predict.

Before reporting the results of this experiment, let me first remind you how strong a phonological activation effect we found for target words like **sheep** in the previous experiment (i.e., the lexical decision latencies for phonological probes like **sheet**). They are given in Figure 7, together with the results for the unrelated test probes (such as **house**) as a comparison.

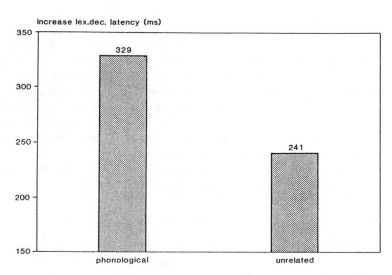

Fig. 7. Increase of lexical decision for probes phonologically related to the target and for unrelated probes.

Now compare this to the phonological activation we found in the present experiment, i.e., for the semantic alternatives (i.e., for probes such as **goal**. These results are presented in Figure 8, together with the results for the unrelated test probes (such as **house**).

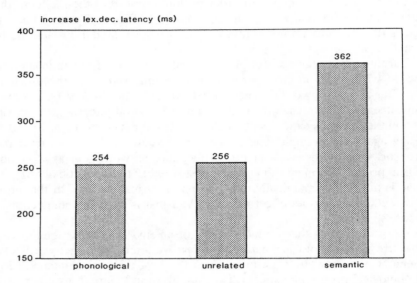

Fig. 8. Increase in lexical decision latency for probes that are phonologically related to a semantic alternative, for unrelated probes, and for probes that are semantic alternatives themselves.

There is not the slightest trace of phonological activation (the result for phonological probes is not different from the result for unrelated probes), contrary to the connectionist predictions. One might, of course object that there was no activation of the semantic alternatives (such as **goat** to start with. But that is not so. Figure 8 also presents the lexical decision latencies for semantic alternatives such as **goat**. There is a highly significant effect here if one compares the results for these semantic alternatives to those for unrelated lexical decision probes (such as **house**).

For a more comprehensive and balanced treatment of the above findings, the reader is kindly referred to Levelt *et al.* (1991).

CONCLUSION

Taken together, the reported results support the modular two-stage notion of lexical access. (Further experimental support for this notion can be found in Schriefers et al. 1990.) An important remaining question is: what could be the biological utility of such a modular architecture for lexical access? The obvious answer is that modularity is nature's protection against error-proneness of a system. The two components of the lexical accessing mechanism have to perform wildly different tasks. Lexical selection involves fast search in a huge lexicon. Phonological encoding involves the creation of a motor program for a single selected lexical item. If these processes were to interact, one would increase mutual interference without obvious functional advantages. Such interference would lead to errors of lexical selection and of phonological encoding. Though errors of these kinds do occur, their rate is astonishingly low for a process so complex and so fast as lexical access. Errors of lexical selection are probably below one per mille selected items, and errors of phonological encoding are even rarer.

To end with a metaphor: if you want to design a reliable car, you better don't connect the action of the brakes to the action of the steering wheel.

NOTES

[1] Apart from myself, the research team involved Herbert Schriefers, Antje Meyer, and Thomas Pechmann.
[2] The model was largely developed by Dirk Vorberg with the assistance of Jaap Havinga.

BIBLIOGRAPHY

Levelt, Willem J. M.: 1989, *Speaking: From Intention to Articulation*, MIT Press, Cambridge, Massachusetts.

Levelt, Willem J. M., Herbert Schriefers, Dirk Vorberg, Antje S. Meyer, Thomas Pechmann and Jaap Havinga: 1991, 'The Time Course of Lexical Access in Speech Production. A Study of Picture Naming', *Psychological Review* **98**, 122—142.

Schriefers, Herbert, Antje S. Meyer and Willem J. M. Levelt: 1990, 'Exploring the Time Course of Lexical Access in Production: Picture-word-interference Studies', *Journal of Memory and Language* **29**, 86—102.

NOTES ON CONTRIBUTORS

Werner Abraham, Department of German, University of Groningen, Groningen, The Netherlands.

Harry Bracken, Department of Philosophy, University of Groningen, Groningen, The Netherlands.

Noam Chomsky, Institute Professor, Massachusetts Institute of Technology, Cambridge, Massachusetts.

Wayne Harbert, Department of Modern Languages and Linguistics, Cornell University, Ithaca, New York.

Celia Jakubowicz, Laboratoire de Psychologie Expérimentale, C.N.R.S., Paris, France.

Shyam Kapur, Department of Linguistics, University of Pennsylvania, Pennsylvania.

Willem Levelt, Max Planck Institute for Psycholinguistics, Nijmegen, The Netherlands.

Barabara Lust, Department of Modern Languages and Linguistics, Cornell University, Ithaca, New York.

Rita Manzini, Department of Linguistics, University College, London, England.

Gita Martohardjono, Department of Modern Languages and Linguistics, Cornell University, Ithaca, New York.

Eric Reuland, Department of Linguistics, Research Institute of Language and Speech OTS, University of Utrecht, The Netherlands.

Thomas Roeper, Department of Linguistics, University of Massachusetts at Amherst, Amherst, Massachusetts.

Jill de Villiers, Department of Linguistics, Smith-College, Massachusetts.

Kenneth Wexler, Department of Brain and Cognitive Sciences, Massachusetts Institute of Technology, Cambridge, Massachusetts.

INDEX OF NAMES

INDEX OF SUBJECTS